WHAT IS SUSTAINABLE

Remembering Our Way Home

Richard Reese

Copyright © Richard Adrian Reese, 2011
All rights reserved

ISBN 978-1466215504

Cover photograph: At the Water's Edge – Piegan.
Photograph by Edward S. Curtis, ©1910.
Edward S. Curtis Collection, Prints and Photographs Division, Library of Congress, LC-USZ62-101262.

http://wildancestors.blogspot.com

TABLE OF CONTENTS

Greetings!   i

PILGRIMAGE TO THE KEWEENAW

Midlife Crisis	1
Finding My Path	4
Springtime Finally Comes	5
A Gentler Way of Living	7
Jewel and Hugo	12
Walter and the Copper People	14
The Anishinabe Blues	21

THE ANCESTORS' JOURNEY

Importance of Ancestors	23
Finding My Roots	24
Welsh Ancestors	26
German "Zipser" Ancestors	34
Norse Ancestors	38
Zooming Out — Seeing With Long Eyes	41

THE GREAT LEAP FORWARD?

In the Beginning — We Are All Related	45
The Vital Importance of Fairies	45
Nomadic Foraging Worked	47
Venus and the Cave Paintings	50
Cultural Evolution	52
Complex Language	53
Ballistic Weapons	54
Are Humans Fatally Flawed?	57
The Curse of Progress	61

FORKS IN THE PATH

The Dark Juju of Stuff	63
The Dawn of McAnimals	66
The First Farmers	76
Exploiting Domesticated Humans	83
Heavy Metal	88
Sacred Forests	92
The Mother of All Forks	98

CHALLENGES FOR AGRICULTURE

Magical Topsoil	115
Annuals & Perennials	117
Watering the Crops	119
Feeding Our Green Relatives	126
Weeds, Pests, and Pathogens	143
Breeding Wonder Crops	153
Junk Food	157
Peak Food	162

SEARCHING FOR SUSTAINABLE FOOD

The Nile	166
Yellow River	168
Slash & Burn in the New Guinea Highlands	171
Old-fashioned Low-impact Organic Agriculture	173
Flooded Paddy Rice Farming	177
The Three Sisters in the Mississippian Era	181
Aquaculture	188
Grass-fed Grazing	194
Miscellaneous Edible Critters	199
Edible Prairies	204
Edible Forests	207
Tikopia — An Island of Sustainability	213
Is Agriculture Necessary?	216

THE TEMPORARY POPULATION BUBBLE

The Hawk and the Squirrel	221
The Reindeer of St. Matthews Island	223
Sacred Man-Eating Predators	224
Dersu the Trapper	225
Sacred Man-Killing Microbes	227
The Brief Golden Age of Antibiotics	230
Reverend Malthus	234
The Population Taboo	237

HEALING OUR WORLDVIEW

A Tale of Two Mental Worlds	246
What is a Worldview?	252
Nazi Hope Fiends	254
Ghost Dancers	257
Ragnarök — Nature Bats Last	260
Benefits of a Shrinking Herd	262
Reevaluating Necessities	266
Healing Ourselves	271
The Prison of "Positive Thinking"	274
Clear Thinking & Deep Awareness	277

HOMECOMING

Marriage to the Land	281
Land of the Copper People	284
Chipmunks & Dragonflies	287
Sacred Migrations	289
Fireflies	291
Deer & Copper People	292
10,000 BC	294
Contented Gwen	296
Always Do That Which Is Best For The Land	298
Keeping the Spirit Fire	298
Thunderstorm	300

HAPPY ENDING

A Great Healing is Unfolding	303
The River Has Long Eyes	305
Fairy Music	305
The Sustainability Paradox	308
Coyote Lessons	311

SELECTED READINGS	**313**
INDEX	**321**
ACKNOWLEDGEMENTS	**329**

Table of Contents

Greetings!

The Earth Crisis is a crazy dance of seven-point-something billion people making a mess of the planet. It is a huge and complex problem with ancient roots. The cure, of course, is a transition to sustainable living, a mode that most cultures have forgotten. What does sustainable mean? A sustainable way of life is one that can survive for many thousands of years without diminishing the ecosystem. Long ago, our ancestors lived in a relatively sustainable way. They spent their lives in a healthy and balanced land, and when they died they left the world in no worse shape than they had found it. This is the one right way to live.

Today, variations of the word *sustainable* can be heard and read dozens of times per day. Beware! Almost always, the word is misused to present a fantasy of purely ersatz sustainability — like saving the world by shopping or magical thinking. The goals of this book are far more dignified and elevated. It will discuss nothing but good old-fashioned fundamentalist sustainability.

Our society prefers not to think about sustainability, because it's so alien to the way we live. It's a shameful and embarrassing subject. Perhaps your parents never sat down with you and discussed it. This is not uncommon. Indeed, it seems like most people in our society have never received "the birds and the bees talk" explaining the mysteries and juicy delights of sustainability — hence, this book, which provides a brief introduction to a very important subject.

We live in an era of big uncomfortable changes, and the future promises more of the same. Our current mode of living is in a process of disintegration, for many reasons. This is good, because it is extremely destructive. Many possible paths lie before us, some are stupid, and some are smart. By definition, unsustainable ways of living can only be temporary. Therefore, in the long run, our two possible destinations are either sustainability or oblivion.

We can move toward sustainability directly and mindfully, or we can stumble through all of the unsustainable alternatives first, causing even more unnecessary destruction. Returning to a sustainable way of life will be a long and turbulent journey, but we have nothing to lose — except for a ridiculous way of life, an abusive relationship with the biosphere, and a highly questionable future for humankind.

The last 200 years have been an era of rapid and extreme change. Living in this hurricane has severed us from our roots. We perceive turbulent, whirlwind change as being the normal condition of life. We have forgotten our past, including an age when our ancestors lived in harmony. We need to remember. We can't possibly know where we are if we don't know where we've been, and we can't be sure of the best path forward when we are lost.

We are hopeless if we don't understand the authentic meaning of sustainability. Dreaming of a sustainable future based on high technology is silly magical thinking. There is no amazing silver bullet solution to the Earth Crisis that lies buried under a big rock somewhere, just waiting to be found. It is naïve to imagine that the generation alive today can set everything right. Healing will take centuries, but all of us are capable of making generous contributions to the process.

In writing this book, I made a deliberate effort to think outside of our standard worldview and its myths. I made an effort to present ideas from a perspective that is more Earth-centered, closer to the perspective of our ancestors who lived sustainably. The path to healing is primarily going to be a head trip. Healing begins when we make a decision to accept responsibility for our lives, and alter our behavior accordingly. Healing begins as we rid our minds of dysfunctional beliefs, and move in the direction of clear thinking. Healing begins when we abandon the madhouse, and remember how to live fully present in reality. The path to healing is ultimately about abandoning industrial society, and remembering who we are, and how to live in balance.

The purpose of this book is to encourage the healing process, to propose important questions, and to contemplate our reality from a different perspective. It is intended to inspire dreamers to embark on voyages of learning, imagination, and creative energy — to inspire pioneers to look for workable paths for our long journey home. The ship is sinking. What is the best route to safe ground? What should we seek? What should we leave behind?

This book is not a catalog of solutions. It does not provide detailed instructions for the correct path home. It does not reveal the one and only Truth. It explores many realms of knowledge, expresses many controversial opinions, and very likely overlooks or misunderstands some issues of importance. Don't trust me. Learn for yourself and think for yourself. One thing is certain: the future demands a radically more intelligent worldview and skill set — and a genuinely healthy future will have little in common with the way we live today.

The four directions of this book are sustainable population, sustainable worldview, sustainable food, and reconnection with our past, our ancestors, and our non-human relatives — the living world. From time to time, I will make appearances, sharing stories from my own quest for a sustainable life.

Best wishes on your own quest. I hope that the next hundred years will be far more intelligent than the last hundred.

I am not a fan of lengthy introductions. Let's go!

PILGRIMAGE TO THE KEWEENAW

Midlife Crisis

It has passed — nine long years in the woods, a sacred journey. Years of growth and learning, pain and healing. An escape from the madness. Living without neckties and alarm clocks and many modern conveniences. Getting closer to nature, closer to my ancestors, closer to the generations yet-to-be-born. A voyage of discovery. Listen. I have a story.

In 1992 I turned 40 and had a midlife crisis. At the time, I was working in Kalamazoo, Michigan, writing computer manuals. I lived in a super-cheap roach-infested ghetto apartment — endless traffic, stinking air. I had cut my expenses to the bone, and was saving like crazy. For many years I had been putting away money to buy a small farm in northern Michigan, a place to live simply, cheaply, self-sufficiently, but so far it was just a dream.

I had steady professional employment, but something was missing. Something was wrong. I needed more. I was planning to leave soon on a two-week vacation. One lucky day the phone rang, and my life took a sharp turn — in the right direction.

It was Thursday, July 16, 1992, maybe 2:00 in the afternoon. Ring! Sitting at the computer in my office. Ring! I was hard at work on some amazingly mindless stuff. Ring! I picked up the phone. Hello. It was Jan, who owned a technical writing firm. She was losing a writer, and needed to hire a new one right away. She offered me nearly twice as much money as I was currently making. I would need to move to Ann Arbor, a town with less heavy industry, and a more intriguing culture.

Same work, nicer town, twice as much money — tough decision, right? I could save even more money, which would allow me to buy a nicer place in the north country. Sure! I'm interested! She called back a few hours later, and we set up a meeting for Sunday.

I went to Ann Arbor and met with her. The business had just moved out of her basement, and into an office plaza. She was rolling in money — shiny new car, fancy clothes, cell phone (an expensive and exotic luxury in those days). She was taking on major new clients. The future looked golden.

I returned to my Kalamazoo ghetto apartment, opened a beer, and for some reason got really depressed. Black metallic roaring inside the skull. Nervous system misfiring and sputtering in a cloud of burnt wiring. Was my intuition trying to convey a subtle message to me? I decided I needed some sleep.

When I got up the next morning, I felt terrible. I so much wanted to move on to a richer and more enjoyable phase of my life. But moving to another big city, and spending even more years writing even more computer manuals did not feel like the path with a heart. It occurred to me that taking this job was maybe a mistake, despite the money. I focused on the concept of mistake for a while, and started to feel better. After an hour or so, it was clear what needed to be done. I called Jan and turned down the job. What a relief!

This was Monday, the start of a two week vacation. I quickly loaded my van and drove north, heading for the Keweenaw Peninsula on Lake Superior. Over the years, I had spent a lot of time exploring northern Michigan. I had come to love the Keweenaw, an old copper mining district, because it was rugged, wild, beautiful, and cheap. This was where I wanted to buy my dream ranch. My passion was the Earth, ecology, history — exploring the Earth Crisis, and ways to deal with it. I wanted to devote more time to what I enjoyed doing — exploring, learning, thinking, writing. I wanted a refuge, and I wanted freedom. It was time to find it.

On Tuesday, I went to the real estate office. On Wednesday, I looked at an old farmhouse just north of Hancock. It was located near the top of a ridge running the length of the Keweenaw Peninsula, a long, narrow spine of rugged land jutting out into Lake Superior. It was a big house on ten acres for $23,000 (equal to my current savings).

There was telephone and electrical service, it was just a mile from the highway, the roof was new, and the outside of the house didn't look bad. The place was probably 100 years old, and it wasn't in perfect condition. But it was livable and cheap!

As I wandered around, I was amazed at the quiet! The place was isolated and secluded. It was about seven miles east of Lake Superior, and ten miles west of Keweenaw Bay (a part of Lake Superior). The grocery store was two miles away. It was essentially just what I was looking for. My enthusiasm grew.

I went back to the realtor, and put an offer on the house. The next morning, Thursday, I learned that my offer had been accepted. On Friday, I wrote a check, closed the sale, and got the keys to the house of my dreams — just eight days after Jan had called me with the lucrative job offer. Buying the house felt right. It was the path with a heart. I was glowing with hope and delight and relief.

The Keweenaw is a gem. Since it is nearly surrounded by Lake Superior, the summer temperatures are usually mild, and the winters are famous for getting 200 to 300 inches of snow. On rare and sacred nights, the northern lights are just gorgeous. There are 800 miles of obsolete railroad lines, built by the mining industry, most of which have been converted to trails. You can spend all day walking in the woods and rarely see another person.

Sixty miles to the southwest, there are 40,000 acres of virgin pine forest at the Porcupine Mountains State Park — an amazing land of large old trees and thundering waterfalls. Out in Lake Superior is the Isle Royale National Wilderness Park, which is a wild island forty miles long and ten miles wide. It has no roads or year round residents, but lots of moose, and a few wolf packs.

I spent the second week of my vacation at my new house, and then drove back to Kalamazoo where I gave my nine months' notice. I needed to work for a while to replenish my finances. I wanted to move north in May, to begin planting fruit trees, berries, and a garden.

I spent the winter getting ready to move. I went through my collection of stuff, and got rid of much of it. I eliminated all frivolous spending — I learned not to carry money with me when I went out for a walk. I also learned to contemplate all new purchases for a minimum of 30 days. I discovered that most purchases were purely impulsive. There are few material impulses that can survive a month of contemplation and reflection. My savings account was quickly replenished.

Finding My Path

During this winter of preparation, I devoted a lot of thought to what I wanted to do in the coming years. I needed to get out of the rat race, and leave the big city far behind. I needed nature and soil and growing things.

One of my main goals was to write a book. For years, I had been studying the Earth Crisis. I was convinced that it was the number one problem confronting humankind today. Learning about the problems, and imagining solutions seemed to be an honorable and meaningful way for me to spend the coming years. I had no wife or children, and thus my material needs were small.

At that point in time, I was frustrated by the damage caused by the Earth Crisis, and by how little was being done to correct it. We knew what the problems were, but the solutions required fundamental change, and substantial participation. Few seemed interested in doing what needed to be done — or even thinking about it. Why not? This made no sense! It seemed to me that the Earth Crisis was completely unnecessary — we understand the problems, there was nothing mysterious about the causes or solutions.

In addition to this frustration, there was also a spiritual dimension emerging in my life. A friend had introduced me to Tom Brown's books on nature and Apache folkways, and I devoured them. I was also hanging out on a computer bulletin board devoted to environmental discussions, where I had illuminating conversations with two Native American gentlemen who had never forgotten the wisdom of their traditions. They had never lost their love and reverence for all Creation — a powerful passion for life that is so foreign to our culture, but so essential for a healthy life.

What seemed to be core to any solution for the Earth Crisis was this sense of connection to all life — reason alone was woefully inadequate to inspire the needed radical change. It felt good to be in contact with a form of spirituality similar to how my northern European ancestors once saw the world. It was like a homing beam — a path to an older consciousness, when the Earth was healthy, and my ancestors knew their proper place in Creation.

So, it was my intention to allow myself to be open to spiritual growth, in hopes of strengthening my connection to the sacred family of life. This path of loving reverence provided a nice counter balance to the chaotic non-stop turbulence of my rational mind. It was a counterbalance to the pain of living with my eyes wide open, heart ablaze.

To survive on the ranch, I planned to make money by doing green writing and selling my homegrown produce — an extremely unrealistic plan, as it turned out. Luckily, merely on a whim, I also decided to keep my main corporate client. This provided just enough money to keep me alive and happy. It worked out perfectly.

Springtime Finally Comes

After the longest nine months in my life, the last day at work finally arrived with the coming of May. I said goodbye to my coworkers, packed my stuff, and went north. Ah! My incredibly vast ten-acre ranch — alder brush, cedar swamp, beaver meadow, apple trees, and recovering forest! Joy! It was overwhelming — to have my own house, owned free and clear, in the Upper Peninsula of Michigan. I couldn't believe my good fortune. Many dream about doing this, but few succeed.

The weather had recently warmed a bit, and most of the long winter's snow had just melted. Only patches remained, here and there. The pond was swollen with new water, the beavers were busy, and the nights were roaring with the deafening high-pitched love chorus of spring peepers. The land was brown and very wet. The air was fresh and moist and rich.

In the Keweenaw, the last frost can be as late as early June. I had four weeks to clear and till and plant the land for the coming summer. I bought a rototiller and hand tools, and ordered trees, plants, and seeds. As a city dude who had been sitting in front of a computer for the last five years, I was out of shape, pudgy, and my hands were very soft.

I planted 400 strawberries, 100 raspberries, 25 blackberries, 10 rhubarbs, 10 currants, 10 elderberries, 25 gooseberries, 10 cranberries, 25 blueberries, 6 kiwis, 10 bush apricots, 10 bush cherries, 10 bush plums, 20 lingon berries, 100 asparagus, 7 cherry trees, 4 pear trees, 3 plum trees, 4 apricot trees, 6 apple trees. Most of these plants died over the coming years — I made a huge and expensive mistake in neglecting to seek advice from the local green thumbs.

I also created a large vegetable garden. I put in green beans, peas, zucchini, broccoli, onions, carrots, beets, turnips, potatoes, cauliflower, chili peppers, celery, herbs, and greens. It was hard work, and the arrival of the mosquitoes and black flies made the process even more uncomfortable. Also, the glacial soil of the Keweenaw is extremely rocky, which really slowed the tilling process — countless ten pound stones had to be dug out. Hundred pound stones had to be pried out, and rolled away.

For the first time in a long time, I felt like I had "a life" — a pleasant sensation. The more weight I lost, the more my pants slipped down. My mind became clear and sharp, and I was remarkably relaxed. I slept like the dead, and awoke clear, calm, refreshed, and happy. This was the life that I had dreamed of. Indeed, my dream had come true.

Eventually, the preparation work and planting was finished, right on schedule, and I was able to relax a bit, and enjoy what I had created. The days were warming up, the leaves were all out, flowers were blooming, birds singing — summer had arrived! It was time to walk and bike and write and explore.

A Gentler Way of Living

One of my goals in moving north was to try to see just how much I could simplify my life, and reduce my consumption. I had already been working on this in Kalamazoo, where I had lived two years without a refrigerator, and my electric bills were down to fifty cents per month (I didn't pay for heat or hot water). On the ranch, my electricity consumption increased to three or four dollars per month. Power was mostly used for a reading light, laptop computer, transistor radio, stove, and water pump.

With bills so small, I was never able to justify the huge cost of putting in a photovoltaic (PV) system. Every six years or so, the PV system's expensive and highly toxic batteries would wear out, and have to be replaced. The system would have taken centuries to pay for itself, if ever. Instead, a healthy dose of conservation made far more sense.

The average American home consumes 1,000 kilowatt hours of electricity per month, while my consumption ranged between 30 and 40 per month — for many years. This simplification required no sacrifice. My quality of life actually improved, as I liberated myself from expensive and frivolous luxuries — instant hot water, automatic climate control, mechanical refrigeration, and so on.

My home came with a refrigerator, but I only plugged it in for a month or two, when my elderly mother stayed with me, and needed to keep her insulin chilled. I had no use for refrigerators, other than as (unplugged) storage cabinets. They made a lot of noise, produced a lot of heat, cost a fortune to run, and their refrigerant chemicals destroy the ozone layer. Without a refrigerator, I shopped often, and bought less. I bought exactly what I was going to eat, and then I ate it. Waste was rare. I had to go out for daily exercise anyway, so visiting the store every day or two was no added burden.

Winters in the Keweenaw were long. The ground was usually covered with snow from early November through most of March. In my house, heat was provided by a new high efficiency wood stove, which burned much cleaner than cheap old-fashioned smoke belchers. I insulated the first floor ceiling, sealed off the upstairs, and slept in the living room during the winters. Each year, I burned about three dump truck loads of wood. In 1993, this cost me $375, and by 2001 it cost $600.

When I installed the woodstove, I removed an antique oil space heater. I didn't want to burn oil, because it is a non-renewable resource. When you burn wood, you acquire your energy locally, keeping the money in your community. Of course, the trees were cut with a chainsaw, hauled with a skidder, split with machine, delivered with a truck — my cheap firewood was produced with cheap oil and industrial technology. In theory, wood is a renewable energy source, but logging is very hard on soils, streams, and forest ecosystems. In practice, the manner in which I heated with wood was harmful to the land. Everything has a price.

My house had a large electric water heater, but I never turned it on. It made no sense to keep 50 gallons of water heated, 24 hours a day, every day of the year. When I needed a gallon of hot water, I simply heated a gallon of water. In the summers, I bathed in the ponds. In the winters, I bathed at a public sauna, which cost three dollars per visit. Later, I built my own sauna.

A sauna doesn't just get you clean, it relaxes you deeply and steam cleans your respiratory system. It was quite thrilling to run naked from the 200° sauna into the fresh biting cold winds of a howling Lake Superior winter storm. Outdoors, I rubbed icy fresh snow on my red steaming skin, and experienced immense pleasure. Then I'd return to the sauna and repeat the process. I always slept well on sauna nights!

In nine years, I used the tub in my bathroom once. I used the toilet maybe five times. It made no sense to dump great fertilizer into a hole in the ground. Poop was gathered in five gallon buckets, and then dumped on the compost pile to break down. Pee, which never contains pathogens, was gathered in separate buckets and applied directly to fruit trees (mixed with some water). The mineral-rich ash from the wood stove was also used as fertilizer

The poop buckets were wonderful portable toilets. You could take one out in the woods, and be in nature while crapping — feasting on the flowers, reveling in the bird music, watching the clouds. In the winter, snow provided a wet and refreshing alternative to toilet paper. In the summer, wet brown leaves made a pleasant wipe. Now, when I go into town, I find crapping indoors to be weird, unpleasant, and unnatural.

In the winter, I shoveled snow by hand instead of using a plow or snow blower. I parked close to the road to minimize my shoveling. Yes, I did have a car. My business required that I drive to work once or twice a year. My annual mileage was much lower than the average American. I didn't visit my friends often because I felt so guilty about driving — it's certainly one of our filthiest bad habits. I much preferred to walk or ride my bike.

I had dumped my television twelve years earlier, an act I never regretted. This freed up countless hours for meaningful activities — reading, writing, walking, visiting. Since I rarely worked, people often worried that I must be suffering from massive boredom. Never once! Oddly, I felt much more time crunched on the ranch than I did working full time in the big city. I never had enough time.

I produced two or three bags of non-recyclable garbage per year, which I took to the waste depot, and paid a dollar per bag to dispose of. This place was also the recycling center, and I took my glass, cans, plastic containers, and magazines there, three or four times a year. Kitchen scraps and yard wastes were dumped on the compost pile beside the garden, which turned it into fabulous fertilizer over the course of two years. I paid attention while shopping, to avoid acquiring wasteful packaging.

White office paper was used on both sides and then recycled at the university. Other scrap paper was used to start fires. I didn't subscribe to the local paper, and only bought it when it was interesting — less than a dozen issues per year. I was careful to give my address only to friends, family, and business contacts, to avoid getting junk mail. I'll never forget the time that I gave ten dollars to an environmental organization — they gave my address to every eco-group in the entire solar system, and all of them barraged me with junk mail for years — many hundreds of pounds of senseless waste!

I washed my clothes at the coin laundry in town maybe four times a year — this was much cheaper than buying and running a machine at home. It's smarter for 400 people to share the same machine, and keep it running regularly — instead of buying 400 machines and using them rarely. Wet clothes were brought home and hung on lines in the shed, or dried by the wood stove. I wore the same clothes many days in a row, to minimize the laundry routine, to extend the lifespan of my clothing, and to save energy and money.

I confess to buying a new power lawnmower during the first summer. The previous owner had kept a large area of the yard cut, and mowing the grass took three or four hours. Well, each year I mowed less and less. Eventually, the lawn was a small square that could be mowed in just a few minutes — a nice place to sit in the shade of Grandmother Oak. In the uncut areas, the wild flowers that grew every summer looked much prettier than the butch cut lawn they had replaced.

Water came from a stone-lined shallow well, maybe 15 feet deep. An electric pump in the basement fed the water into the house. During dry summers, the well would get low, and it wasn't possible to water the garden. In the winter, the pump would always freeze and quit working for months. So, I'd fill buckets with snow and bring them indoors to melt.

In 1997, the pump froze and cracked open. It cost $450 to replace it. In 2001, it froze again. This time, I didn't replace it. I switched to rainwater harvesting in the summer months. A local bakery often put out free food-grade plastic buckets, which I brought home and collected. When it rained I set the buckets around the roof edges and gathered my water. The water from rain and snow was clearer and tasted better than the well water.

It was an educational experience producing my water without machines. I learned how to use every drop of water two or three times. For example: (1) use it for cooking, then (2) use it to soak dirty dishes, and then (3) use it to water the fruit trees. On average, I consumed about four gallons per day — for cooking, cleaning, and bathing. Minimizing waste was a source of satisfaction for me. On visits to the big city, I'd listen to people taking 20-minute showers, and it would drive me out of my mind — the waste — and they'd do this every single day. Oh my God!

When you spend weeks splitting, hauling, and stacking firewood, you don't waste it, because you have respect for all of the hard work invested in it. You don't keep the house at 80° — you put on shoes and socks, an extra shirt, a sweater, a warm cap — and then you can stay comfortable until the temperature drops down into the 40's. The sense of coldness is largely in the head, and formed by habit. It can be reprogrammed to be comfortable at much lower temperatures. One thing that struck European colonists about primitive people was that they seemed to be impervious to cold — the !Kung bushmen would happily sleep naked outdoors on nights when it got down into the 30's. Learning this simple trick can save a fortune in heating bills — and it's surprisingly easy to learn.

When you grow your own food, you understand how much time and effort is required to produce each onion, potato, and zucchini. Consequently, you develop a heightened sensitivity to waste — and you go out of your way to avoid it. On visits to "normal" households, it's heartbreaking to see how much food an average family tosses in the garbage.

When you pay attention to your power consumption, and deliberately try to keep it low, you develop practical and thrifty habits for living. It's stunning to see how much power the average family wastes — lights on in empty rooms, televisions on without viewers, yard lights, space heaters, air conditioning, clothes dryers.

When you learn the pleasure of walking, and notice how good it makes you feel, you lose a lot of interest in driving — an expensive, stressful, and unpleasant habit. You begin to have regrets about all of the frivolous driving you've done in the past.

If everyone were required to live for a year in a manner similar to how I did, the world would be a much better place. People would pay far more attention to how they lived. There would be so much less waste!

My Keweenaw friends had similar lifestyles, and some lived more lightly than I did. Living simply was not a matter of sacrifice or discomfort. At the ranch, I enjoyed nine rich and stimulating years while consuming far less than the average American. Cutting my spending reduced my need for money, which reduced my need for working, which liberated huge amounts of time for interesting and satisfying activities. Over the years, I worked an average of less than six hours per week for money. When you avoid wasting time and money, your life becomes richer, deeper, and happier.

Jewel and Hugo

During the first summer on the ranch, I gradually unpacked my stuff. At the same time, I had to move out the belongings of a dead lady, Jewel, the previous owner. I often felt her spirit watching me with a generous degree of disapproval, as this renegade eco-nut hermit moved into her orderly home. I felt like an intruder for the first few months. Her husband Hugo, who died in 1968, was a farmer and a blacksmith at the copper mine. I often felt his presence, too.

Within a few weeks of living there, I became really aware of Jewel and Hugo's incredible thrift. They never threw out a single button, bit of thread, or scrap of fabric. I gathered up a big jar full of buttons, and a half-dozen large boxes of fabric scraps. Had Jewel lived longer, they would have been made into rugs and quilts.

In the basement were boxes and boxes of scrap wood to be burned in the kitchen range. They picked up every fallen branch and used it. Pruned limbs were gathered and saved. Broken furniture that couldn't be fixed was cut up and put in a box. So were the segments of old fence posts that weren't rotten, and boards and cedar shingles from fallen outbuildings. Wooden boxes were brought home from the mine, taken apart, and stacked in neat piles.

Every nail and screw was saved. A bent nail could be straightened and used. It was an object of value, not to be wasted. I found a hand-woven rug, four feet by six feet; made out of plastic grocery bags, trash bags, and bread bags. Why discard stuff that can be made into something useful?

When the old galvanized washtub developed a leak, it was patched, not discarded. There were old steel tractor wheels in the yard where the cracked rims were wrapped with sheet metal and then wound with wire.

I was amazed by how carefully they bought, how carefully they used, how much they preserved, and how little they wasted. Dollars were never spent foolishly here. Nothing was bought for more than it was worth, or sold for less. Whatever could be produced at home, was. Never was anything, with any conceivable type of usefulness, ever discarded. Compared to them, I felt like an incredibly wasteful person. Jewel and Hugo showed me a higher standard for simple living. I will never forget their lesson.

The summer passed, the frost came, and then the snow. At this time, I started the biggest learning process of my life. I began reading — ecological history, anthropology, folklore, archeology, environmental ethics, and on and on. I wanted to understand how the Earth Crisis began, how it grew, and what could be done to resolve it.

I never stopped reading — books, magazines, websites — and I took detailed notes on everything. I joined Internet mailing lists, and had long discussions with many different thinkers and dreamers — pilgrims of all flavors, from every corner of the world. I studied through the end of 1993, and then all of 1994, and then 95, 96, 97... Now it's 2011, and I'm still at it, more than ever.

Walter and the Copper People

The Anishinabe are the indigenous people of the Keweenaw, and much of the Great White North. They are also called the Chippewas or Ojibwas. One of the tribe's most sacred possessions is the Protect the Earth staff. Worthy individuals are chosen for the honorary role of carrying the staff. Walter Bresette was the man who had this honor and responsibility. He lived at the Red Cliff reservation in Wisconsin.

In 1997, Walter was invited to Michigan Technological University (MTU) to teach a course called "Alternative Views on the Environment." My friend Vern Simula had been talking to Walter, and mentioned that I was a local thinker with more than a few alternative views. Walter agreed to give me two hours to present my views to the class. One day, Vern brought Walter over to meet me. I described this event to a mailing list devoted to discussing Daniel Quinn's book *Ishmael*:

> Today, I had a wonderful experience. A genuine, full-blooded, fundamentalist animist walked through my door, sat down, and started telling stories. Walter is an Anishinabe activist and storyteller, a being of big love, radiant with spiritual vitality. His world seems so alive, intense, and meaningful. It seems like he's solidly plugged in to the planetary mainline, while my connection is weak, sputtering, sporadic.
>
> The first story he told was about the White Pine Mine, in northern Michigan, which had recently shut down operations. An Anishinabe delegation of maybe 20 people came to the mouth of the mine, to perform a traditional ceremony. Their

purpose was to help heal the injuries that had been inflicted upon the land, during many decades of destructive greed. The ceremony let the spirits of the mine know that the horror was over, the machinery would not come back, and that the healing process could now begin.

At the mouth of the mine, a sacred fire was built, and there was drumming, songs, and prayers. Some of the elders were wary of standing too close to the opening because such awesomely powerful (and possibly dangerous) spirits dwelled within that dark, gaping, painful wound in the Earth.

When the ceremony was over, the people moved indoors, to the mine's conference room, for a feast. Elders smudged the room with the smoke of burning sage, blankets were laid on the floor, and foods were set out — venison, wild rice, fry bread. Before people ate, prayers were offered, the sacred pipe was passed, stories were told, and gifts given. It was anything but an ordinary evening in corporate America!

At Walter's invitation, the president of the mining corporation had flown in from Toronto. Although they had been in conflict for many months, the two had come to respect one another. Both shared touching and humorous stories. It was clear to all that bonds of brotherly love had been formed. Walter was a man who radiated with an aura of warm love, and this big love had the power to touch and soften the hearts of his adversaries. His love was good!

After we had lunch, I showed Walter a large, green, beautifully shaped piece of float copper (float copper is copper that has been smoothed by passing glaciers). To the Anishinabe, all of Creation is alive, sacred, and related — everything in Creation is a being with spiritual power. This includes every form of flora and fauna. This also includes things that modern scientists myopically refer to as inanimate, like copper rocks. In fact, copper rocks are especially sacred to the Anishinabe, and they affectionately and respectfully refer to these ancient and powerful red metal spirit beings as the *copper people*.

Walter took the heavy green stone in his hands, and gasped with amazement and delight. The two of them went off, into the next room, and sat down together by the wood stove. Walter bowed his head, and he and the copper spirits spent a long time in deep and sacred communication. It was an awesome and moving experience to observe.

Walter had been fighting mines for years. Now he held in his hands a sacred member of the copper people, a beautiful being who had been spared from the crushers, smelters, and factories. Watching the two together was a deeply moving experience! Brother embracing brother. Love and radiance filling the room.

Ladies and gentlemen, this profound love for the Earth — for the spirits of all beings, animate and inanimate, past, present, and future — is what is missing in our discussions, in our books, in our culture, our religions, our schools, in each of our lives. If our hearts are filled and brimming with love for the community that surrounds us — the *entire* community — then we would not need to spend our time in these discussions — we would need to spend our time in prayer, in song, and in sharing. The Earth has the answers we are seeking. It is our sacred obligation to seek, to ask, and — most important — to listen!

When we can remember and rekindle that love — the overwhelming, boundless, bottomless love that our ancestors had for the Earth, all these questions that are bouncing around in cyberspace will become pointless and silly. When our hearts are hot and strong and radiant with that love, we will cease causing harm — we will do what is right, proper, and good. We do not destroy what we love. Love is the answer. Community is the answer.

October 30 was the day I spoke to the class. It was also my 45th birthday. That night, Walter and a dozen friends came to my home for a ceremony — a night I'll never forget.

Walter was one of the most spiritually alive people I have ever met. He was fascinated by my house, because I had hundreds of copper people on display. One of my favorite hobbies was finding copper at old mine sites. He spent the night, because he wanted to sleep and dream in a sanctuary of copper spirits. The next day, I wrote:

> On Thor's day there was a gathering at my home, in the land of the copper people. A feast for the dead. A celebration. A ceremony of the Anishinabe People, the indigenous people of this land — led by a beautiful being named Walter.
>
> I am bound by my honor not to reveal what took place in that gathering, but I can tell you that it was good. It was healing. It was strong. It will be remembered for long years. It was a time of softness, gentleness, peace.
>
> On the morning of Halloween, I got up, built a fire, and made coffee. In a cloud of tobacco smoke, Walter and I talked. I asked Walter for feedback on my talk of the previous day. My words to those students were not charged with tension, fear, and anxiety — but with directness, with sincerity, with power. But some said that my words had stirred fear. This was not my intent. My motive had been to inspire thought, concern, awakening, action — to inspire seeking and questioning, learning and communication, healing and transformation, gentleness and peace.
>
> I wanted to start a spreading fire that would lick the spirits of the masses — and race across the land, cleansing and purifying, burning and renewing, killing and giving birth. But Walter told me that the mystery he was pursuing was not one that required many people — just a few — maybe ten or twenty.

He wasn't playing the savior game. He had no plans to lead a revolution.

I had asked the students a question: why did most of the early civilizations in the lands that we call the United States collapse — even before the invasion of the whites? Was it a shortage of firewood (they had no axes made of the iron people)? Was it disease? Did the terrible new technology of the bow and arrow enable over-hunting? Or was it a spontaneous collective act of intelligence, of healing, of gentleness and peace?

Walter reflected for a moment, and then he told me a story. He told me of a legend he had heard of, the legend of a people who lived south of the border, up in the mountains. In these mountains are ruins, the ruins of an ancient civilization. These people told Walter that one-day their ancestors abruptly packed up, walked out of the city, and closed the great gates behind them — never to return again. They had mindfully and deliberately abandoned the civilization that had been making their lives so unbearably miserable — what a powerful and healing dream!

I had told the students that our culture and its schools only talks to one brain, the left brain, the brain of linear thought, reason, rationality, analysis. Our right brain withers, shrinks, gathers dust. Our right brain is the brain of creativity, of vision, of prayer, of spiritual power. I suggested to the students that since we all have two brains, maybe we ought to use *both* of them!

We ethereal Ishmaelites, connected by electrons passing through a network of wires made of copper people, think and speak mostly with the language of the left brain. Argument and rebuttal. Point and counter-point. Here is what Walter said to me: "If reason and rationality had *real* power, *real* medicine, our problems would have been solved long, long, *long* ago."

Walter described his two lives to me. One life is public — he works for native rights, for peace, for the environment. He travels around the world. He has met many famous people, and many unknown people who have important stories to tell. His other life is spiritual, a sacred quest, a pilgrimage, a search for answers to great mysteries.

Walter's sacred life reminded me of the stories of Carlos Castaneda and Don Juan. It was a world of spirits, powers, mysteries, visions, prophesies, stolen fires, frozen objects, ancient beings, lost scrolls, forgotten caves — a world of seeking, of questioning, of discovering, of remembering.

I have lived for 45 years mostly by using my left brain. To my mind, Walter's sacred world was 100% irrational, mysterious, incomprehensible, purely and absolutely nonsensical. Yet, it stirred inside me profound feelings, forces, spirits, something important. There was much that my people had forgotten, much that we needed to remember.

I suggested that Walter speak of his sacred life to the students, because it was of such great importance. He laughed! "If I told them those stories, they'd all start following me! I'd never be able to get rid of them!" He said that many white people are looking for a leader, someone to give them instructions, someone to follow! They don't know how to conduct their own lives, to find their own paths, to live their own visions.

Suddenly, Walter got up, picked up his bags, declined an offer of breakfast, hugged me, and walked out the door. I just sat still. I spent the entire day doing nothing. I didn't read anything. I didn't turn on my computer. I didn't return phone messages. The radio stayed quiet.

This Halloween day was like no other day that I have lived in all of my 45 years. It was like a dream. I was completely relaxed, comfortable, content, satisfied, fulfilled. For the entire day, hour after blessed hour, I was a being of immense and total peace, happiness, and quiet. There is one last thing to tell here. There are some people who now know me by a new name — an Anishinabe name. In English, this name means Copper Man.

When his course at MTU concluded, Walter went home. The excitement faded, I returned to my studies, the winter snows began, and life went back to normal mode. Our meeting was an important event in my life. It was a precious gift to spend time with a person from an indigenous culture — a person with a powerful living spirituality, and a big radiant loving presence. Walter was unlike any person I had ever met before, and he became a role model for me.

I had spent my entire life around people with rational analytical civilized minds, and Walter clearly came from a far more balanced and wholesome reality. There was a huge gulf between us because the cultures that raised us were so different. Being with him made me feel so clumsy, ignorant, immature — childlike.

Walter radiated an intense reverence and respect for the land, and everything that dwelt here — everything was alive and sacred. This place was the land of his ancestors, his home, his place of worship, where he belonged. The land and his people were one. My ancestors came from many places on the far side of the ocean, and I am a rootless tumbleweed. I have no tribe, I have no healthy culture, I have no wise elders, and I have no home — there is no place where I belong. I am lost and alone.

Walter Bresette passed to the other side on February 21, 1999.

The Anishinabe Blues

The Upper Peninsula of Michigan is known as the U.P. (and its residents are therefore known as *Yoopers*). Fur traders moved into the region around 1650, but significant European settlement in the UP didn't occur until the mid-19th century. The invader's culture was tremendously different from the Anishinabe culture — like night and day, like good and evil, like war and peace.

Nobody knows exactly how long the Anishinabe have been living in the UP, but they have been there many centuries (they have traditions describing an ancient migration from the Atlantic coast). We do know that when the white explorers arrived, the UP was a healthy, thriving, vibrant wilderness. There were few people, endless forests of giant white pines, abundant wildlife, and astonishing numbers of birds and fish. The air, water, and soil were fresh and clean and pure.

Following the explorers came wave after wave of miners. The beaver miners arrived in the mid-1600s, and by 1800 the beavers were approaching extinction. Then came the mineral miners. Copper and iron mining started around 1840, and both peaked around 1920. Then came the tree miners, who stripped away the ancient forests in a 50 year logging binge — the UP was mostly stumps by 1900. Farmers settled on the richest stump lands, and commenced to mine the soil. The fish miners wiped out billions of whitefish by the 1890s, then switched to lake trout. Fish mining peaked around 1915, and then fell into a permanent decline.

These miners were not skilled at clear thinking. Whites moving into new regions never failed to be astounded by the wealth of the land, and the unbelievable abundance of wildlife. They believed that these "resources" were essentially infinite. The iron deposits were so rich that they could be mined for eternity. There were so many white pines that it would be impossible to cut them all down. The schools of whitefish were so vast that all the fishermen in the world could never catch even half of them. But, the whitefish were wiped out, the white pines were cut down, the copper mines and iron mines were emptied and closed. Amazingly, the time it took to destroy each of these treasures was essentially one human lifespan — 60 to 80 years.

In the nearly 400 years that European invaders have lived in the Great Lakes region, we have never put down roots and blended into the land. We've always been like moon explorers, connected by long cords to distant industrial complexes, in order to survive. The fur trappers needed their traps, guns, knives. The loggers needed their saws, axes, chains. The miners needed their drills, hoists, crushers. The farmers needed their plows, fences, tractors. The fishers needed their boats, engines, lifts. The consumers needed their cars, TVs, petroleum.

We've never been able to cut our link to industrial civilization, and live off the land, or blend into the ecosystem, our family. Just like the moon explorers, we zoomed in, planted a flag, dumped a bunch of trash, and then zoomed out. We thought like colonists, and we lived like colonists — when this place got too screwed up, we'd pack up and go somewhere else, exterminate the indigenous folks, and repeat the same mistakes.

Clearly, the settlers had a very different relationship with the land than the Anishinabe had. The settlers simply devoured everything having monetary value, turned it into cash, and then moved on to other ecological disasters. Why did they live in such a destructive manner? What made them think so differently?

A long, long time ago, Europeans once lived in a manner similar to the Anishinabe. But they changed. Something happened. Something dark and painful — civilization. Europe was invaded by people from the Fertile Crescent, and the ancient European way of life was destroyed. Their wisdom, knowledge, songs, and stories were lost. The cultures of Europe became more and more dysfunctional. Later, Europeans spread this madness to other continents, and destroyed these lands, too. Why? This is one of the most important stories of all. This is something we need to remember, and think about, and talk about.

THE ANCESTORS' JOURNEY

In the first chapter, I told you a bit about who I am. In this chapter, I'm going to devote more attention to my ancestors.

There is a tradition, practiced by some Native American cultures, that I am fond of. When two people meet for the first time, the process of introduction can take hours. The first person describes his name, his tribe, his clan, the features of the land he belongs to, the journey of his life. Then he describes his parents in a similar manner, and then his grandparents. When he is finished, the other person introduces himself. This process is richer, deeper, and more meaningful than: "Hi, my name is Richard Reese, and I'm a technical writer from Kalamazoo."

Importance of Ancestors

In the old days, the notion of "family" included the living, the dead, and the generations yet-to-be-born. The presence of the ancestral spirits was accepted as a normal part of everyday life, and they were welcomed and venerated. Ancestors were consulted for wise advice. They served as guardians, watching over their living descendants, and providing assistance — like we watch over children. Ancestors also served as cultural police — they punished, menaced, or haunted those who violated the taboos. They drove some people crazy, or even killed them. So, they had a sweet and beneficial aspect, and a scary one.

The word *ancestor* still has a holy resonance to it. Ancestors have been venerated since ancient times, because of the important role they played in ensuring the continuity of the culture. While alive, they learned time-proven wisdom from their elders, preserved it, and passed it along to the young generations. They kept the culture alive and well.

Everyone had a sacred obligation to become a good ancestor to the generations yet-to-be-born. This system worked wonderfully, as long as the functional time-proven culture remained strong and healthy. When the culture became sick, the sick culture was passed along, and the community deteriorated or died. It is important that the younger generations understand that these times of sickness are not normal, and do not resemble our ancient path.

Finding My Roots

I'd like to provide you with a brief introduction to my ancestors, because the story of my family is the Earth Crisis in a nutshell. I was born in 1952 in Oakland county Michigan, not far from Detroit. I grew up in a brand new subdivision of 96 homes, built on George Ogg's farm. It was surrounded by forests, wetlands, lakes, farms, horse stables, and orchards. I spent my childhood playing in the woods and swamps, and this was a precious gift, I now realize — feeling at home in the natural world, having regular contact with wild animals. Many never have this experience — the ancient, normal, healthy experience of being a living human in a living land. Disconnection from nature is a big part of our problem.

I graduated from college in 1974, and then spent the following years bouncing from job to job and state to state. My desire was to live an honorable life, but our culture is confusing, and the moral and ethical path was not obvious. Many tempting voices sang sweet songs to lure me down destructive paths. Where were the good paths?

My life has been a sacred pilgrimage in search of meaning and integrity. Who am I? Why am I here? Where am I going? I began studying the history of my people, and I was fascinated by the indigenous culture of pre-Christian Europe. In 1977, I decided to explore my family tree, and spent a couple summers in Europe, wandering and searching. I met relatives in Norway and Wales, and stood on my ancient ancestral bloodlands. It was a deeply meaningful experience.

In the mid-1980s, I began tuning in on environmental issues, and discovered that this was another creepy realm. The more I studied it, the creepier it got. The Earth Crisis is not just a modern problem, it has deep roots, going back many centuries. My ancestors have struggled for generations against its stormy seas.

And so, without further ado, I'd like you to meet my ancestors. Here are the clans of my mother and father:

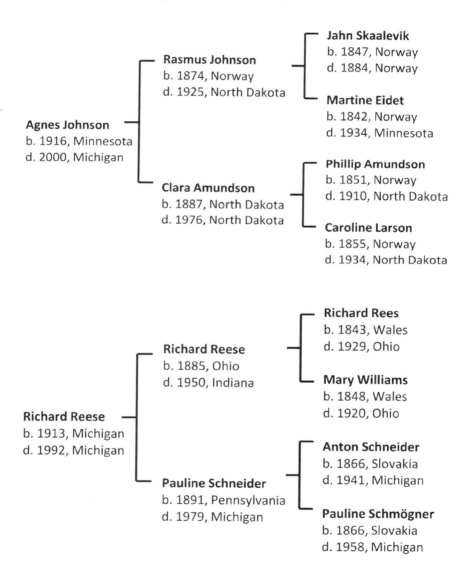

Welsh Ancestors

I am one-fourth Welsh, and my Rees ancestors came from the hamlet of Cwmbelan in the parish of Llangurig, a hilly, remote, and sparsely populated region not far from the English border. About 80% of Welsh lands are devoted to the agricultural sector, but because of the hilly terrain and the marginal soils, little farming is done in Wales. Instead, the dominant activity is raising sheep, which provide meat, wool, and hides. Today, Wales is mostly pastureland with few trees. The hills look like a gang of skinheads.

It's hard to imagine that the land was once covered with abundant forests 2,100 years ago, prior to the arrival of the Iron Age. It's hard to imagine the rivers thrashing with countless salmon. It's hard to imagine the abundance of wildlife that once thrived in this land — aurochs (wild cattle), brown bears, cave bears, cave lions, elk (moose), grey wolves, Irish elks, Eurasian lynx, reindeer, wisent (buffalo), wolverines, wooly mammoths, wooly rhinoceros, and great auks. All of these species went extinct in Britain, many by human hands.

It's hard to imagine ancient tribal people in Wales living low-impact lives in a vibrant forest filled with wildlife, but they certainly existed, and their blood runs in my veins. These ancestors talk to me in my dreams. It saddens them to see how we suffer in the modern world. Their message is simply this: come home! We miss you! Let the land heal!

Sheep and Textiles

In the early 1800s, my Welsh ancestors were cottage artisans who made woolen cloth. They were handloom weavers. In their day, they enjoyed what amounted to a stable middle class way of life.

In Cwmbelan there was a water-powered factory that made flannel cloth. Water power is a renewable form of energy, and it emits no greenhouse gasses. The Cwmbelan factory was built beside a waterfall on a small stream called Nant Cwm-Belan. The factory did nothing that artisans could not do in their cottages, but it manufactured products with less labor input, which could be sold for less.

For most of the human journey, muscle power was the primary form of work energy. But during the medieval era, there was a power revolution — waterwheels and windmills came into widespread use, taking over many muscle-powered tasks. The new technology was immensely clever, but it wasn't perfect. Winds varied in intensity, and windmills worked best in locations with strong and steady winds. The volume of water in flowing streams also varied, and waterwheels only worked beside streams with strong flows.

The desire for a dependable source of power that could be used anywhere inspired the imaginations of devious minds. After some tinkering, trial, and error, they unveiled the coal-fired steam engine, a diabolical monstrosity that was unsustainable in every way. But it sure was profitable! A 100 horsepower steam engine could power 50,000 spindles, and spin as much yarn as 200,000 men. Profits soared, and the Industrial Revolution was born. Huge new mills were built in the English counties of Yorkshire and Lancashire, and northern England became the center of the textile industry.

This system excelled at producing low-quality products at low prices, using unskilled workers who worked long hours for slave wages. Following the 1845 famine in Ireland, hundreds of thousands of Irish fled to England and Wales, and they were delighted to find any work at all. Living in rags and eating potatoes was more fun than starving to death in Ireland. The greedy capitalists welcomed them with open arms.

The Luddite Rebellion

Cottage artisans performed skilled labor that was satisfying, and gave them a sense of pride and dignity. This work generated an adequate income, which was not the norm for many working people at that time. With the coming of mills and power looms, 500,000 handloom weavers lost their livelihood to the unsustainable new technology. They lost their future, and the futures of their children and grandchildren. Next stop: poverty.

Today, it's hard for us to imagine how hard this transition blindsided countless rural communities, in a land where the wool business was the primary business. There were no social safety nets at that time, no welfare offices, so starvation was poverty's shadow. Many cottage artisans were forced to leave their homes, move to filthy cities, and work for wages that were miserably low.

Llanidloes is two miles down the road from Cwmbelan. The *Municipal History of Llanidloes* provides abundant information about the rich and powerful, and little about the working people. But one short sentence shouts the painful truth: "About 1800 the introduction of machinery doubled the Poor Rates."

In Lancashire, riots began in 1768, followed by machine breaking in 1779. The natives were not happy. To make life even more miserable, the early 1800s added the challenges of food shortages and sharply rising food prices. The anger level grew until it exploded with the Luddite Rebellion of 1811 to 1816. Large numbers of uppity people ran about in the night smashing machines, and the general public was quite sympathetic. The government sent in a massive force of 12,000 soldiers to crush the Ludds.

This uprising represented a serious challenge to the rapidly-growing monster of heartless steamroller capitalism — a powerful new economic system that had the mind of a two-year-old ("Mine! Give it to me!"). Communism, which came later, was not the antidote. Both systems were possessed by a self-defeating compulsion to pursue planet-killing industrial civilization. Neither system had a spiritual connection to the living planet, and the generations yet-to-be-born, of all species.

The Luddites are commonly presented as morons who foolishly opposed brilliant advances in wonderful technology. Actually, they were intelligent and hard-working people who opposed cold-hearted greed, mass unemployment, and starvation. Sending letters to their corrupt government did nothing. Would peaceful protests have accomplished anything? Only the rich could vote. The masses voted by riot and rebellion, and they voted often.

I have a feeling that most people everywhere would respond in a similar manner to such ruthless and extreme injustice. The power of the Luddite Rebellion eventually led to the birth of the Labor Party. We all have an inner Luddite, and it is precious. It gives us the strength to stand defiantly in opposition to greed and exploitation.

Today, when discussing the issues of modern living, a common conversational prefix is "I'm no Luddite, but..." This prefix is always followed by a legitimate complaint about technological civilization. For example, "I'm no Luddite, but I think that our use of pesticides is taking a big toll on our health." It's like an apology for wanting to say something important and intelligent. It's completely unhip to be a critic of progress, no matter how unhealthy and destructive the progress really is.

The coal-powered steam engines and power looms were not just socially destructive, they also had a heavy ecological footprint. They required the existence of a steel industry, which required the existence of coal and iron mining industries. They increased the harm that civilization caused to the ecosphere, while providing no long-term benefits, and creating many long-term problems. Indeed, they helped to greatly accelerate the pace of the snowballing catastrophe that continues to this day.

At the same time — very importantly — we should never forget that the cottage artisans did not represent a sustainable way of life. (Could you imagine spending your entire life hunched over a spinning wheel?). Over time, the ever-growing wool industry turned the vast ancient forests of Wales into pastures. It exterminated a number of sheep-eating predator species. It radically changed the Welsh ecosystem by simplifying it, by eliminating balance and an abundance of diversity. My wild salmon-eating ancestors are horrified by what has happened to their once-magnificent land. It's a heart-breaking thing to see!

The Tithe Commutation Survey of 1836 estimated the land usage of every region for revenue purposes. Here is the breakdown for the parish of Llangurig, where my family lived:

Land Use	Acres
Arable land (cropland)	3,125
Meadow or pasture	9, 375
Common land	37,000
Woodland	100
Glebe (church) land	4
Total	49,604

The size of the parish was 49,604 acres, but there were only 100 acres of surviving woodland. The forest was on the brink of extinction! Also, take note of the 37,000 acres of common land, for this leads us to our next subject.

Enclosure of the Commons

Land ownership and usage varied by region, but the general rule was that a few wealthy people owned most of the land. Peasants rarely owned land, most were tenants. The tenancy agreement provided the peasant with a cottage and gave him the privilege of using specific cropland and pasture. So, on a large estate, some land was assigned to specific tenants, and the rest was commons — land that all tenants on the estate had the right to use, limited by rules and restrictions (i.e., no tree cutting, no hunting, no fishing, no picking berries, etc.).

We have seen that there were 37,000 acres of commons in Llangurig parish in 1836, almost 75% of the parish. But in the 1875 book, *The History of the Parish of Llangurig*, we learn that since the 1836 survey, "large quantities of the common land have been enclosed." Enclosure is the process that eliminates access to land that was formerly commons, and assigns the right of use to a specific person. This reduced the resources that each tenant had access to, and it often forced poor tenants off the estate — and many times this was the objective of enclosure, because peasants were often a messy, annoying, and unreliable source of income.

Allow me to introduce you to the verb "unpeople." The enclosure movement unpeopled thousands of towns and villages. The abandoned buildings were cleared away (except for church structures), and the old communities ceased to exist. Today, we call them DMVs (deserted medieval villages).

The enclosure movement started around the 13th century, and was pretty well finished by 1850. It paralleled the growth of the wool industry, which was very profitable for the lords and ladies of England, Scotland, and Wales. It all boiled down to money. Generally speaking, an estate inhabited by a large number of poor subsistence farmers generated far less rent than a handful of shepherds raising thousands of sheep. So, the commons were enclosed, the peasants were booted off the land, and they migrated to filthy cities to be exploited by the merciless capitalists. Enclosure meant more poverty, more sheep, fewer trees, and more greenhouse gases spewed by the coal-burning textile mills. It stimulated the growth of cities and industry. It did not benefit my ancestors, the common folk, nor did it allow the land to heal.

Here's a memorable enclosure story. The Isle of Rum lies off the west coast of Scotland. Humans have lived there for maybe 9,000 years. They survived by hunting, herding, fishing, and subsistence farming. The Lord of Rum was disappointed that the island generated just £300 of income per year, so he enclosed the land in 1826. The 350 residents were put on boats and sent to Nova Scotia, Canada, and they were heartbroken about leaving the only home they had ever known. The nine hamlets of subsistence farms were replaced with one sheep ranch that covered the entire island — 8,000 black faced sheep were brought in. Rent payments shot up to £800 per year, which put a big smile on the Lord's face.

The Exodus

My great-grandfather Richard Rees was born in Cwmbelan in 1843. Eight months later, his father died of "decline" at the age of 23, leaving a widow and three sons. Richard was born near the dawn of an era of decline for the parish of Llangurig. In the 1841 census, the parish population hit a peak of 1,957 residents. This was followed by decades of steady decline. By 1930 there were just 908 residents. The last school in the parish was closed in 2008, having just eight students.

In Llangurig, the golden age for cottage artisans was over, and the textile industry was migrating to the mills in big cities. The golden age for the local lead-mining industry was over. Enclosure was eliminating a number of tenements, liberating many peasants from a means for survival. The future did not look bright for a young widow with three sons. They had two options: they could move to northern England and work in the mills, or move to southern Wales and work in the iron and coal mines. Both places were hopping, but neither was a worker's paradise.

They decided to go south. Around 1853 they moved to Dowlais, Glamorganshire. The boys became iron miners, and mom ran the Green Dragon pub. It was a hellish life — hard work, wee wages, filthy air, raw sewage everywhere, and tainted water that killed thousands. In 1863, they packed up and moved to Pittsburgh, Pennsylvania — an equally charming steel town (formerly the home of the Iroquois people). Two years later, they bought a boat and rowed down the Ohio River to Pomeroy, Ohio (formerly the home of the Delaware people).

Richard's future wife, Mary Anne Williams, was born in 1848 in Tredegar, about ten miles east of Dowlais. Her father was a miner at the notorious Sirhowy Ironworks. In those days the River Sirhowy was nicknamed the Black River (it matched the color of the coal smoke air). Because of the filthy conditions, cholera regularly came to visit (1832, 1833, 1849, 1866), killing both rich and poor. Cholera was known as the "King of Terrors." The epidemic of 1849 was especially severe.

Legends tell of families being alive and well in the morning, and all dead by evening. Doctors had no cure. Mary Ann, her parents, and three siblings survived the epidemic. In 1851, they packed up and moved to Pennsylvania. By 1860, the Williams had moved to Minersville, Ohio, and in 1869 Mary Ann married Richard Rees.

Richard and Mary Ann raised a family while bouncing from coal mine to coal mine in Ohio. They started in Minersville (1869), then Thomastown (1874), then Pigeon Run (1876) where four of their five children died in an influenza epidemic, then Shawnee (1879), then Hemlock (1900), then Crooksville (1910), then Columbus (1920) where they retired. Richard lived to be 86, and he worked in the mines for 65 years — ten in the old country, two in Pennsylvania, and 53 in Ohio!

Today, the Dowlais Ironworks is gone, and its huge open pit is known as the biggest hole in Europe. Nearby is one of Europe's biggest hazardous waste landfills. The Sirhowy Ironworks is gone. Poverty is widespread in southern Wales. The steel industry in Pittsburgh is gone. The coal industry is gone in Minersville, Thomastown, Pigeon Run, Shawnee, Hemlock, and Crooksville. The industrial way of life is like a string of firecrackers: boom, bust, boom, bust, boom, bust…

My grandfather, Richard Reese, was born in Shawnee in 1885. He moved to Detroit around 1912, and spent the rest of his life there (formerly the home of the Ottawa, Huron, and Potawatomi people). He worked as a presser in a pottery factory, a machinist in an auto factory, a strawberry farmer, a conductor for the Detroit Street Railway, the proprietor of a confectionary store and gas station, and then a street side kiosk.

My father Richard was born in 1913, and spent his life within 30 miles of Detroit, working as an accountant. My father and grandfather both witnessed the boom years of Detroit — an era of skyrocketing economic growth, explosive population growth, ferocious urban sprawl, and colossal environmental destruction. The party peaked around 1960, and today Detroit is a city in ruins, gradually returning to forest — 40 square miles of the city (28%) is abandoned properties.

In 1952, I was born in Oakland County, Michigan. In 1950 the population in the county was 396,000. In 2010 the number of people had *tripled*, to 1,202,000. I spent my first 18 years in West Bloomfield Township. In 1950, there were 8,720 people in West Bloomfield. In 2010, there were 63,634!

When I was a boy the county was rustic, lots of wild places and open spaces. Today the county is completely developed — everything paved, people everywhere, impossible traffic jams. There is nothing left of the lush and beautiful land of my childhood. My home has been destroyed by a population explosion. The region is nearing the end of its boom phase. We know what phase comes next.

We have to be thankful that there is not enough petroleum, fresh water, or cropland to keep this diabolical growth continuing for another 50 years. If it did, there would be nothing left.

I sometimes dream of what my life would have been like if Wales had not been overrun by invaders from an unhealthy culture. I imagine living in a beautiful ancient forest along the River Severn, spending my days hunting for deer and fishing for salmon; sitting beside a warm fire under a starry sky, singing sacred songs and telling ancient stories. It would have been a good life — living in a healthy culture with a healthy future.

German "Zipser" Ancestors

Long ago, some of my wild ancestors lived along a river in the forests of Germany, probably the Rhine. They were buffalo (wisent) hunters and salmon eaters. These ancestors talk to me in my dreams. It saddens them to see how we suffer in the modern world. Their message is simply this: return to the path of balance! We miss you! Let the land heal!

Salmon once thrived in the rivers of Europe, Japan, Siberia, and the east and west coasts of North America. In Europe, Ice Age cave artists painted this sacred fish. More frequently, they painted images of sacred buffalo.

Julius Caesar wrote one of the earliest descriptions of the tribal Germans in 51 BC. Most of the tribes were sedentary cattle herders whose diet majored in meat, milk, and cheese. They hunted, and they gathered nuts, berries, and fruit. Some farmed a bit, but their fields were depleted after one season of use. Better soils remained safely protected by endless heavy forests. At that time, the tribes had few metal tools. Caesar said that you could travel 60 days in the vast Hercynian forest and not find the edge of it. It covered the mountains from the Rhine to Romania, and was home to bison and aurochs.

A population explosion blew my ancestors out of Germany, east to Hungary, starting in the 12th century. A number of significant changes came together to provide a sharp boost to food production.

- A new and improved tree-felling ax appeared in the 10th century, which made it easier to clear forests. In this era, the cleric Adam of Bremen wrote that the highlands of Germany were covered with frightfully dense woodlands.

- The moldboard plow had come into widespread use in northern Europe. These heavy-duty plows had metal edges that sliced through the turf, and then turned the soil over, providing thorough cultivation. This breakthrough enabled the tilling of heavy clay soils, which were rich in nutrients. Previously, it was not possible for farmers to use these soils. The moldboard plow led to a rapid expansion of cropland.

- Horses replaced oxen as the primary work animal. This was made possible by the invention of horseshoes, and the development of harnesses designed for horses. Horses worked faster, so more work could be done each day. Since they were quicker than oxen, they could also work fields farther from the village.

- In northern Europe, the three-field crop rotation method was replacing the two-field system. In the two-field system, half of the land was fallow (unplanted) each year, to partially restore its fertility. In the three-field system, only one-third of the cropland was left fallow. The result of this shift was an increase in the amount of land that was producing food each season.

- At the same time, it was a period of significant global warming. During the Medieval Warm Period (800 to 1300 AD), temperatures were warmer than in the previous 8,000 years (and warmer than the toasty 20th century, too). This climate change benefitted agricultural productivity in Europe. It also allowed cropland to expand into higher elevations, which had previously been too chilly for farming. Of course, this expansion required deforestation. In the 5th century, about 95% of Europe was forested, but by the early 17th century forest cover had been reduced to 20%.

Because of these changes, the population of Europe doubled between 1000 and 1300. Many Germans packed up and moved east, across the Elbe river, to colonize "underdeveloped" lands. This was known as the *Ostkolonisation*, the colonization of the East.

My people called themselves the Zipsers, because they lived in the Zips region of the Carpathian Mountains in Hungary (which is today the Spiš region of Slovakia). They mined gold, silver, and copper until the lodes petered out. Then they mined coal and iron. This led to a bustling iron forge business that manufactured farm tools, nails, and chain. My great-grandmother, Pauline, lived in Svedlar. My great-grandfather, Anton, lived 15 miles away in Gelnica. Both towns are on the Hnilec River.

Cholera and typhus frequently came to thin the crowds in the villages of the Hnilec Valley. In 1710, an epidemic killed three-quarters of the residents of Svedlar. I did a survey of 100 Gelnica death records from 1840 through 1873, and noted the age of death. Twenty-two died before their first birthday. Another 38 died before their fifth birthday. Two-thirds died before reaching 25. The average age of death was 18.8 years. The ages of the oldest five deaths were 70, 70, 71, 73, and 74. My great-grandmother was just four feet, nine inches tall (1.44 meters).

The iron forge business did well until the 1860s, when the Industrial Revolution drove it into the ground. Big city factories mass-produced the same products and sold them for less. Many of the metal workers and miners packed up and moved away. The railroad arrived in 1871, and it sped the decline by making it easier for skilled workers to move to where the jobs were. By the 1920s, the Hnilec Valley was nicknamed Hunger Valley. In 1945, the Red Army drove the Germans out of the region. Today, Svedlar has just 1,993 residents, half of them Gypsies. Unemployment is about 50%.

My great-grandparents decided to leave the Zips. Anton cruised to New York on the Main in 1886, and Pauline took the Columbia in 1889. They married in Manhattan in 1891, and then moved to Pittsburgh, where they raised six children (all of whom survived to adulthood!). Anton worked at the Jones and Locklin Steel Company. Pittsburgh was known as the Iron City. It was the center of America's steel-making industry, and it was in its rapid growth phase. By 1916, they had moved to Detroit, where the economy and the population grew for the rest of their lives. They wanted to be with their oldest daughter, Pauline, my grandmother, who had married Richard Reese in 1912.

In the land of the Zipsers, Anton and Pauline grew up in isolated small towns surrounded by forested mountains. Their ancestors had lived in the same place for centuries. Everyone was related to everyone. It was rare to see a stranger. Coming to America was an overwhelming experience — their lives shifted into a frantic fast forward. New York, Pittsburgh, Detroit — massive bustling places, filled with busy strangers — noise, crowding, pollution, crime.

Anton became a mean alcoholic, and he died at Eloise, an insane asylum in Detroit. Pauline lived to be 91. I met her in 1957, and shook the hand of someone born in 1866. Their house on Bessemore Avenue is now gone. Their once-thriving neighborhood is now a burned out ghetto. When they were born, the vast virgin white pine forests of northern Michigan still stood, and millions of buffalo still ran free on the western plains.

In the 1930s, one of their sons achieved brief fame when he was arrested, along with 15 other members of the Black Legion, suspected of plotting to murder a newspaper publisher. The Black Legion was a terrorist hate group with 60,000 to 100,000 members in Michigan, Indiana, Illinois, and Ohio. They hated Jews, Catholics, Communists, blacks, union organizers, and welfare system bureaucrats and recipients. They bore a resemblance to the fascist groups rising in Europe at that time. One of the schemes the Black Legion contemplated was delivering typhoid-contaminated dairy products to Jewish neighborhoods.

Norse Ancestors

My mother's family came from three rural regions in Norway. Hordaland is on the west coast of Norway, south of Bergen, a land of many rocky islands, few people, and abundant salmon. Buskerud is in the mountainous forests of central Norway, an isolated land with few people. Hadeland is 30 miles north of Oslo, in a fertile valley where there are many farms.

Norway is at the northern edge of Europe, and it has a cold climate. Most of the terrain is hilly or mountainous. Only 3% to 4% of the land is suitable for agriculture. These factors have proven to be a blessing, because they have slowed the rate of destruction. Norway has not been extensively deforested, plowed, urbanized, and industrialized. Much of the land remains wild and beautiful. Norway's limited agricultural productivity has caused social strains during population spurts.

The Viking era (750-1050) was driven by population stresses, and many moved away to colonize other lands. The Black Death arrived in Bergen in 1349. A number of historians believe that the Black Death hit Norway exceptionally hard. Up to two-thirds of the people died, and entire communities were depopulated and abandoned. Farms and villages returned to forest. The recovery process took 200 years.

One of my ancestral villages is Gran in Hadeland. Gran is not far from Grinaker, where there is an ancient stone church. Legend says that several decades after the plague, a small boy discovered the church in the forest. Its existence had been completely forgotten. A similar story describes a hunter deep in the woods. His arrow missed its target but hit the bell of a forgotten church. The hunter looked inside and found a bear hibernating before the altar.

For many years, population growth was slowed by epidemics, famines, and war. Then suddenly, around 1720; the population exploded. The standard explanation for this explosion was "peace, vaccines, and potatoes," but I think that the cause was more complicated. Here are estimates of Norwegian population:

Year	Population
1600	200,000
1700	450,000
1800	880,000
1900	2,100,000*

* Not counting the 500,000 who emigrated

Potatoes arrived in 1758, and they enabled farmers to produce three times more calories per acre than wheat, rye, or oats — using far less labor. The widespread adaptation of potatoes was a gradual process, but by 1810 potatoes had become a popular crop in Norway. They certainly made a substantial contribution to the later stages of the population explosion. Farms became able to feed more people, and people who were better nourished had stronger immune systems and were less likely to die from disease.

In 1798, the cowpox vaccine was introduced, and it reduced smallpox deaths over the next 100 years. By this time, most people were resistant to the bubonic plague — the last major epidemic was in

London in 1666. Municipal sanitation improvements were not a significant factor, because most Norwegians still lived in rural areas, few lived in cities.

The "peace, vaccines, and potatoes" explanation is missing one huge factor: grain imports. Norway exported fish and minerals, but their primary export was lumber. A major import was grain — by the 1820s, 40% of the grain consumed in Norway was imported.

I have relatives at the Gulden farm in Gran, Hadeland. Ancient bronze artifacts have been found in burial mounds at this farm. It's possible that my ancestors have lived here for over 2,000 years. Standing on this land was a powerful experience for me. The effects of the population explosion are evident in the history of Gulden. In 1801, this farm had nine households and 59 people. In 1865, there were ten households and 71 people. In 1977 there was one household and three people.

In 1862, the United States passed the Homestead Act, which made cheap land available. The mass emigration of Norwegians began in the mid-1860s. In the earlier Viking era, growing population drove the Norse to conquer and settle the coastal regions of Ireland, England, Scotland, France, Iceland, Greenland, and Russia. This new population explosion drove the Norse to conquer and settle Wisconsin, Iowa, Minnesota, and the Dakota Territories. My mother's family emigrated in 1870 and 1872, and eventually settled in eastern North Dakota, where they built sod homes, plowed the prairie, and planted wheat (formerly the home of the Lakota people, and vast herds of buffalo). During World War II, my mother worked in Milwaukee, where she met my father.

Norway was blind-sided by the age of colonialism. From the European perspective, the globe doubled in size around 1500, and Europe became the center of the global economy. In the colonies, forests were rapidly cleared to expand cropland area on fertile virgin soils. This led to a sharp upturn in food production. Europe was flooded with cheap imported food, and Norwegian farmers had a hard time competing against it.

At the same time, European farmers also increased their food production when they introduced highly productive New World crops: corn and potatoes. Thus, Europe was awash with cheap population fuel. The population of Europe doubled between 1750 and 1850. One estimate is that 55 million people left Europe between 1830 and World War II.

Zooming Out — Seeing With Long Eyes

Mountaintops are often regarded to be sacred places because you can rise up above the town or forest and see things that are far away — you can see with *long eyes* — you can take in much more than the ordinary perspective allows. From the mountaintop of the ancestors, it is possible to see time with long eyes, to observe changes that have taken place over generations and centuries, to observe the path we have taken, and to foresee where we are going. There is great power in being able to see with long eyes. It is good. You cannot engage in clear thinking without having long eyes.

My Norwegian ancestors fed their increasing number of mouths by trading lumber for grain — essentially, they were feeding many kids with sacred trees. Likewise, my Welsh ancestors in the woolen industry fed their babies by turning forests into sheep pastures, and trading wool for food. My mining and metal working ancestors fed their babies by turning coal and iron into edible calories — they were feeding kids with sacred stones. In other words, population grew by depleting and diminishing ecosystems — by mining forests, fish, topsoil, and minerals. Population grew at the expense of the generations yet-to-be-born, of all species. This mining era was not guided by clear thinking.

In the journey of my life, I have lived in several lands where the indigenous people did not maintain their way of life by depleting the ecosystem. The Anishinabe, Yurok, Ohlone, Takelma, and Kalapuya people seem to have enjoyed a relatively stable way of life, and a stable population, for thousands of years. Their villages were not fortified with wooden palisades to protect them from belligerent neighbors (palisades, castles, and forts are structures built by people suffering from conflicts rooted in overpopulation).

They did not feed their kids with trees, stones, or petroleum. They fed their kids with wild foods from their land: fish, nuts, berries, meat, greens, and vegetables. They did not enslave plants and animals. They did not plow or deforest or mine. They did not suffer from civilized diseases like bubonic plague, smallpox, cholera, measles, and typhus. Famine was rare or unknown. They had a functional and coherent culture and worldview, and most of them had healthy, meaningful, enjoyable lives.

If we look back far enough, my wild ancestors in Norway, Wales, and Germany also lived like this. *Everybody's* ancestors once lived like this. Hunting and gathering was the normal way of life for almost the entire human journey. They had a sustainable way of life. They had respect and reverence for the land. They were comfortably integrated into the family of life. Their culture and lifestyle worked, and worked well.

Meanwhile, our modern civilization continues to slide farther and farther from this ideal, despite each new century of amazing technological and intellectual progress. Our wild ancestors were not ignorant, primitive, hideous beasts. I like to imagine that they were typically prosperous, successful, healthy, and content.

But my European ancestors did not remain on the path of stability and balance. They were invaded and conquered by people from the east who had learned the dark arts of enslaving plants, animals, and minerals. The invasion of the domesticators gave birth to centuries of chaos and instability. Population grew explosively, despite the explosive growth of warfare, contagious diseases, emigration, and famine. The forests were destroyed. The wildlife was destroyed. The predominant subject in the history of Europe is a bloody stew of wars, warriors, battles, massacres, and empires. Underlying all of the chaos is an ever-growing war on the ecosystem. Can you foresee where this is heading? If you were a gray-haired elder with long eyes and a clear mind, what would you recommend?

When my wild ancestors in Europe were "discovered" by civilized invaders from the east, there were no anthropologists to document the indigenous culture, before it got erased. But I have found one ancient document that provided a glimpse into their world. In 98 AD, the Roman historian Tacitus wrote *Germania*. It briefly described the nomadic foragers who happily lived in the vast unbroken forests of what is now Finland. These people were called the Fennians.

He said that the Fennians dressed in skins and slept on the ground. They had no horses, homes, or proper (metal) weapons — just bows and arrows. The Fenni were superior to the farming people, because they enjoyed a pleasant and leisurely way of life. Safe and secure in their ancient forest home, they feared neither men nor gods. According to Tacitus, they were such contented people that they had no need to pray for anything whatsoever — because they had everything they needed. Life doesn't get any better than that.

Following the Industrial Revolution, my people suffered from turbulent change, after being uprooted from lands where they had lived for centuries, in stable communities. Many became industrial nomads, and spread out across the planet. I have had addresses in nine states, and my extended family has spread out across most of the 50 states. We live in houses, but we no longer have homes, places where we belong, close to where our ancestors are buried.

I have always been envious of the Anishinabe. They still remember the instructions that the Creator gave them. They still remember their songs and stories. They still have wise elders who provide guidance and healing. Being a mongrel with European blood, living in North America, I feel disconnected.

Walter Bresette told me that the ancestral knowledge, the wisdom of my people, was "accessible." He didn't explain how. But I believe this, even if I don't understand it well at the moment. I do, in fact, have a direct linkage to the spirits of my wild ancestors. Through my blood, they remain alive. When I sing, I hear their voices. When I dream, they tell me stories. When the night winds blow, I smell the smoke from their fires.

OK! So now you know a bit about me and my ancestors. Hopefully, this will help you to better understand the following voyage of ideas and stories.

THE GREAT LEAP FORWARD?

In the Beginning — We Are All Related

Once upon a time, long, long ago, the Earth was created. Nobody is exactly sure when. The planet was rock and dust, fire and ice, water and air. Then, some kind of magic juju happened (nobody is exactly sure how), which combined these raw elementals into the first living beings — wee little single-celled creatures, smaller than the eye can see. Some think that this might have happened close to four billion years ago.

Then, millions and millions of years passed, a whole bunch of evolution happened, and these wee beings transformed into millions and millions of species, of every imaginable size, shape, and color. Indeed, all living things — the oak and shark, mosquito and snake, gorilla and clover, ant and whale — share *a common ancestor* in these original wee ones. *We are all related* — one big sacred family — everything that flies, swims, walks, or blooms. This is an important truth to remember. All life is one. Everything is family.

The Vital Importance of Fairies

In the stories of ancient Europe, the earliest creatures were the giants, beings older than time. They were very big, very strong, and had voracious appetites. We can still see them today, covered with soil and trees, sleeping peacefully — they are the mountains. When they awaken, they can become fearsome — they cause earthquakes, tidal waves, thunder, and volcanoes.

Later came the little people — wee folk, fairies, sprites, brownies, kobolds — they had many names. The little people were small, lively, and incredibly intelligent. They lived on wild foods — nuts, roots, meat, and fruit. They never tilled the soil, nor herded animals. Hunting gave them great delight. The primary passions of the little people were music, song, and dance. They were a joyful and carefree people.

Jacob Grimm described their disposition with two delightful words: "inexhaustible cheerfulness."

Fairies lived in Fairyland, the wild places unspoiled by plows, axes, roads, and dwellings. The wee folk were the spirits of the land. Every plant, tree, animal, fish, insect, mountain, and body of water had an alive and vital spirit. Life in Fairyland was sacred, intimate, and beautiful.

And then, one fateful day, the big people appeared. Big people can walk across a land and neither see the spirits nor hear their joyful music. The bumbling clumsiness of the big people brought endless grief to the hearts of the wee folk. The wee ones hated the introduction of grazing animals, and the horrid noise of tinkling sheep bells. They shunned our farm-grown food, because it was so unhealthy; eating it hastened death. They detested our lumbering, our ripping of the Mother's flesh with sharp plows, our weird fascination with numbers and counting, and the blackest art of all — metal making.

With time, the growing madness became just too much for the little people and, clan by clan, they tearfully said good-bye to the lands that they so deeply loved and cared for, packed up their belongings, and set off on long journeys to unknown places.

The stories of the fairies were tales passed on from an earlier era, one older than the age of farmers. The wee folk were not only symbols of the spirits of the land; they were also symbols of the hunters and gatherers — symbols of an ancient age when people lived in harmony with nature, an age of song and dance and joy. With the departure of the fairies, precious sacred lands became disenchanted, and the ecosystem became a warehouse of resources to be plundered and destroyed for personal enrichment.

Some see the healing of the Earth as an act of restoration — bringing back Fairyland to every place. Magic and beauty. Song and dance. Health and balance. Peace and celebration.

Nomadic Foraging Worked

Scientists tell the story of our origins a bit differently. Around three million years ago, the first two-legged *hominids* appeared. We were peculiar animals, because we walked upright on two legs, like birds, but had no wings or feathers. In the stories of many indigenous cultures, hominids were seen as being the newest creatures, and as being a bit clumsy at the survival game — almost childlike. In these tales, other animals often helped us, or taught us useful skills, because they felt sorry for us.

For our first 2,990,000 years, or so, we lived in relative harmony with the Earth (compared to the present generation). We weren't always perfect, though. During the last 40,000 years or so, as we migrated away from our primal African home, we sometimes had problems when entering new regions — especially with critters that moved slower than we did, and with critters who had no instinctive fear of us. We hunted some species to extinction.

Humans altered the composition of a number of ecosystems by periodically burning off the brushy vegetation. We did this to keep wooded areas more open, and to increase the size of open meadows and prairies. This eliminated hiding places for man-eating predators, provided a better habitat for game animals, and made it easier for us to hunt them. It also drove away pesky biting insects.

In *The Ecological Indian*, Shepard Krech wrote that these deliberate fires sometimes got too big, too fast, too hot. Sometimes a thousand animals were killed by a fire. Sometimes the fire damaged the soil, and the vegetation did not grow back.

Because of the activities of Stone Age people, there are some prairies that were once forests, a new desert or two, and dozens of species were driven to extinction.

Until about 10,000 years ago, almost all humans everywhere were nomadic foragers. This way of life worked well. Compared to our modern mode of living, it was vastly less destructive; it left far fewer scars on the land, and was much better integrated with the surrounding ecosystem. There is abundant evidence that nomadic foragers were healthier and happier than people living in or near modern societies.

The music of the Pygmies is often sung in beautiful, enthusiastic ten-part harmonies. These songs celebrate the sacredness of the forest. The purpose of their songs is to keep the forest awake, and to keep the forest happy. They are among the finest love songs in the world.

In about 2500 BC, Egyptian explorers came across the Pygmies. The report to the Pharaoh noted that these jungle people sang to the forest, and danced for it. They have been doing this continuously for four and a half thousand years, and probably much, much longer. While countless civilizations have come and gone, the Pygmies have continued singing in their sacred forests. The Pygmies had a culture that worked. For almost all of human history, successful cultures were the norm.

The anthropologist Colin Turnbull wrote that cooperation was at the core of the Pygmy way of life — and this included their singing, which never included solos. Their songs were a harmony of joyfully cooperating voices. Self-focused individualism was unknown.

All around the world, the successful cultures of these nomadic people *generally* had a number of aspects in common:

- They believed in formless ethereal great spirits, but not in gods and goddesses with human names, faces, and personalities.

- They believed that all life was sacred, interconnected, and related.

- They believed that every rock, stream, plant, and animal had power, spirit, holiness, consciousness, wisdom, and the ability to communicate.

- They believed that all members of the family of life were equal — the whale, the dandelion, the human, the ant. Each had special gifts and powers, and they often taught humans important things. A Kuyukon elder once said: "Any animal knows way more than you do."

- They believed that they had to live in balance with the source of their existence, and carefully limited their numbers in order to avoid the painful problems of excess.

- They did not engage in commerce, or the sale of services. Sharing was the cardinal rule — no one starved while another feasted.

- They believed that personal belongings could become poisonous, if not kept in constant circulation with mandatory gift giving.

- They lived in small bands of ten to thirty people, which allowed everyone to maintain a state of close friendship with all in the clan.

- They believed that everyone in the clan was equal — decisions were made by communal discussion and consensus. They had no judges, chiefs, police, or preachers.

- They treated women with fairness and respect — far better than civilized cultures did.

- They had the most leisurely lifestyle in all of human history. When they were hungry, they went out and found food. Otherwise, they visited, socialized, celebrated, explored, did craftwork, or relaxed.

- They had shamans, herbalists, and other healers who attended to mental or spiritual imbalances before they could develop into physical illnesses.

- They believed that the harmony of the land and the harmony of the clan were closely linked — if the clan got out of balance, so would the land, and vice versa.

- They put the community first. Self-centeredness was not taught, encouraged, or tolerated. People with big heads were relentlessly mocked and teased until the swelling went away and they returned to a state of health and cooperation.

It's fascinating to observe what many modern folks do on our vacations. We journey to the wilderness and camp with our families. We hike, fish, hunt, and swim. We sit around campfires, feasting and telling stories. This is how we spend our one week of freedom, the greatest week of the year — living like wild people, attempting to find our lost inner cave man.

Civilized people have historically looked down on the nomadic foragers. We call them savages and barbarians. We believe that their lives were nasty, brutish, and short. Almost everyone believes this, with one exception — anthropologists — people who have actually had direct face-to-face contact with simpler people. Certainly our wild ancestors were not angels, and they were not perfect in every way, but there is much that they could teach us. For example, they enjoyed a way of life that didn't smash the ecosystem. This would be good to learn.

Venus and the Cave Paintings

For most of human history, our mental and technical abilities changed slowly. The author Tim Flannery noted that, as late as 100,000 years ago, visiting space aliens wouldn't have seen us hominids as anything other than ordinary animals. Jared Diamond wrote that early humans were little different from chimps. We were nothing special, and would not have stood out in a wild community. We were integrated with the family of life, and in balance with it. We left the world in no worse shape than when we found it.

Then, during the last Ice Age, a transition occurred that threatened this ancient state of balance. By 40,000 years ago, hominids *did* stand out from the crowd of wild critters. We were carving Venus figurines, decorating our bodies, and making flutes and rattles. We were painting pictures on stone walls. Some of these cave paintings have survived. Something happened to us. Something shifted us out of our traditional patterns.

A number of respected thinkers describe this shift as the *Great Leap Forward* (which should not be confused with Mao Zedong's Great Leap Forward, in which 20 to 40 million Chinese died of starvation around 1960). In terms of technology, it's an appropriate name. Prior to the great leap, our tool-making skills advanced at a snail's pace. Following the leap, the evolution of our tool-making went into fast forward — eventually opening the doors to agriculture, civilization, industry, high technology, and all the rest.

On the other hand, from the perspective of ecological history, the Great Leap Forward is a name that is comically inappropriate, because this shift also opened the floodgates to the creation of the millions of problems that comprise the Earth Crisis, and pose a serious threat to our future. We acquired a diabolical magic that gave us the power to destroy life on Earth.

To the modern mindset, technology critic George Basalla was guilty of heresy when he wrote that fire, axes, or wheels are not necessities for life. Plants and animals grow quite well and thrive in the absence of farmers, and raw foods are quite nourishing — agriculture and cooking are not necessary. How have chimpanzees managed to survive without using these crucial advances?

The world of technology perceives the Golden Age to be in the future, because everything is getting better all the time. The planet's ecosystem perceives the Golden Age to be in the past, before the emergence of agriculture and civilization. In a heavyweight match between technology and Mother Nature, I would always bet everything on Nature.

Cultural Evolution

Information is passed from one generation to the next in two ways: genetic information and cultural information. Genetic information lies within the realm of sperm and eggs (hardware). Cultural information is the knowledge we learn from others after being born (software). Both types of information evolve. Genetic information evolves slowly, while cultural information can evolve at a head-spinning velocity. Today, cultural information can become obsolete in just a few years. Are you old enough to remember Mighty Mouse, Microsoft Bob, Iron Butterfly, or Betamax?

The Great Leap Forward was the dawn of an era in which cultural information evolved more quickly, and became more complex. Humans were no longer just evolutionary creatures, we became revolutionary creatures. We began to generate and transmit far more cultural information than any other animal.

Paul and Anne Ehrlich have written a lot about cultural evolution. One time when Paul was in northern Canada, visiting with indigenous hunter-gatherers, he was surprised to observe that every person essentially had a complete understanding of the tribe's entire collection of cultural information. Everyone knew how to tan a hide, clean a fish, or weave a basket.

On the other hand, no modern American knows even one-millionth of our culture's information. Someone can get a degree from Stanford but never learn a thing about science. If you spent your entire life reading books at a university library, you'd learn only a tiny portion of our cultural information.

A primary reason why I wrote this book was to provide a thumbnail sketch of ecological history, with an emphasis on what worked and what didn't. Many people know little about this, and it's a very important subject, especially for the centuries that lie before us.

The Great Leap Forward was not the official beginning of the Earth Crisis. It wasn't until the megafauna extinctions and the dawn of agriculture that some cultures began causing serious injuries to their ecosystems. Some cultures remained essentially sustainable into the 20th century, and a handful still survive today. But industrial civilization would not have been possible without the mental abilities that we developed during the Great Leap.

An obvious question is: What caused the Great Leap Forward? Why did cultural evolution slip into fast-forward? Many experts believe that complex language played a primary role in this transition.

Complex Language

Complex language is a powerful skill. It has enhanced our ability to transfer learned information from one generation to the next, eliminating the need for each new generation to start from scratch and relearn the basics via a long and complicated process of trial and error and error and error. It provides us with thousands of years of accumulated knowledge — a substantial advantage in the survival game. As our skills with complex language increased, humans became more and more unusual in the animal world.

Of course other animals have intelligence, and can think, remember, predict, strategize, cooperate, use tools, communicate, and so on. But humans have obviously developed their intelligence and communication skills in a manner that is radically different from other animals. Whether the benefits of these peculiarities outweigh the drawbacks is an open question.

Take a moment to contemplate the advantages that speaking humans had over wordless, grunting, howling, screeching humans. Words promoted complex thinking, complex communication, and complex social relationships. Words provided a huge advantage in the survival game. Every night when we returned from our daily wandering, we would gather at the campfire and everyone would report the latest news from different regions of the countryside (i.e., "buffalo by the river"), discuss it, and then plan our next activities.

On the downside, this power could make us too clever for our own britches. Take modern society, for example. We sophisticated civilized folks have been taught many self-destructive skills and ideas, and complex language, combined with reading and writing, made it easy to pass this self-destructive information from one generation to the next, and to every corner of the world.

On the other hand, because of complex language, we have never had a better understanding of how our self-destructive civilization is ruining our life support system.

But then, even though we understand with a high degree of precision many of our worst mistakes, we continue to repeat the same mistakes, and make new mistakes that we know are even more destructive. Our faulty beliefs and misguided values have deep roots and powerful momentum. This can get confusing — trying to comprehend human history. We couldn't have gotten as messed up as we are today without complex language. Nonetheless, any form of remedial action is going to involve complex language.

Ballistic Weapons

Ballistic weapons have had a major influence on our history. Rock-throwing, a unique human skill, has killed millions of animals over the ages, including other humans. The oldest spears date back somewhere between 250,000 and 400,000 years. The bow and arrow existed in Eurasia 10,000 to 30,000 years ago, but didn't make it to some parts of the New World until 1,000 years ago. Ballistic weapons led to major changes in the human way of life:

- Weapons made us better able to deter attacks from man-eating predators and other humans, thereby reducing our death rate.

- Weapons increased our ability to hunt effectively, thereby increasing our food supply.

- Weapons created a greater need for humans to communicate, cooperate, strategize, and share. Organized teams did a better job of repelling threats and hunting game.

- Weapons made homicide less risky and easier — victims could be killed from a distance, by a hidden assailant, without direct hand-to-hand contact, even if they were much bigger and stronger. This increased the efficiency of warfare.
- Weapons spurred technological innovation, because better tools increased the odds of survival.

For most of us, man-eating predators are no longer a normal part of daily life, so we tend not to think about them. Since we came down from the trees, humans have been highly vulnerable to predation — we weren't fast, we weren't strong, we didn't have big horns, antlers, or tusks. We couldn't even outrun an elderly half-lame leopard. In fact, we were so easy to kill that it's surprising that humans have managed to survive. If we had patiently waited around for evolution to make us fast, strong, and horny, we would have vanished long ago.

Humans have never enjoyed being helpless prey, but this is what we were when we came down from the trees. Evolution had not designed us to have an easy life as ground-dwellers at the forest's edge. Cleverness was the primary reason for our survival (although today our cleverness is destroying us). People who used sharp sticks survived better than the empty-handed. People who cooperated with others survived better than loners. By stimulating our cleverness, man-eating predators certainly played a central role in our "Great Leap Forward."

For much of our journey, humans were primarily foragers and scavengers. Like our chimpanzee cousins, we did hunt, but not with great efficiency. The reason for this was simple: we weren't physically well equipped to be effective hunters — compared to natural-born predators like the wolf, grizzly bear, or jaguar. An empty-handed unarmed human had little chance of ever taking down an elk, duck, or squirrel. Ballistic weapons made it possible for humans to become far more productive hunters. This gave us a profound advantage at the game of survival.

The power of weapons was enhanced by combining it with the power of teamwork. The need for teamwork encouraged the improvement of conflict resolution skills. The ideal was that all disputes should be resolved by bedtime. When a dispute could not be resolved, despite repeated efforts, one of the troublemakers might be sent to live with another clan for a while. A functional clan required a high degree of emotional harmony — and song, dance, storytelling, and ritual were important medicines for re-harmonizing a ruffled group.

A group of humans possessing ballistic weapons, who knew how to cooperate and kill, set the stage for the dawn of war. The anthropological evidence clearly shows that nomadic foragers were generally much less violent than sedentary people, who lived in fixed locations. This suggests that humans are not fundamentally violent by nature. Violence seems to be more related to culture, and driven by stresses like population density, territorial threats, resource shortages, and big-headed leaders. Nomadic foragers like the Pygmies and !Kung bushmen were able to use ballistic weapons for hunting, while rarely spilling human blood.

Around 11,000 BC, stone-tipped weapons came into common use. The earliest stone heads are known as *Clovis* points. These new super-sharp points could penetrate the thick hides of large animals. So, humans could now kill species that had previously been impossible (or foolishly dangerous) to hunt.

These stone-tipped spears have caused a loud decades-long argument among scientists. The dominant faction supports Paul Martin's *Pleistocene overkill* hypothesis, which believes that overhunting was responsible for the rapid extinction of numerous large mammals (megafauna) in North and South America. Others blame climate (a comet strike?), disease, or a mixture of all three. They point out that some toads and mollusks also went extinct in this era.

The Sioux scholar, Vine Deloria, ridiculed the overkill theory. He felt that it implied the stupidity of Native American hunting people. In his book *Red Earth, White Lies*, he calls our attention to the mountains of megafauna bones (mostly mammoths and rhinos) found in northern Siberia and Alaska. Hundreds of thousands of mammoth skeletons provided many tons of tusks for a thriving ivory industry that started in the 1770's. These animals were not killed by hunters, they were victims of a mysterious catastrophe — a huge natural event that the overkill crowd does not acknowledge or explain.

It is popular for misanthropes to present the Pleistocene overkill hypothesis as absolute proof that humankind is fatally flawed. They purport that we, as a species, are so inherently stupid that our only possible destination is self-extinction. Therefore, our fate is already sealed, nothing really matters, and there is no reason for caring about anything. I disagree. Misanthropes tend to suffer from a lack of imagination.

Are Humans Fatally Flawed?

Once upon a time, humans were no more notable than chimpanzees. Then the Great Leap Forward happened, and we went through some major changes. Our communication, organizational, and tool-making skills became more powerful. This transition eventually opened the doors to agriculture, which further multiplied our powers. Agriculture eventually opened the doors to the possibility of modern industrial society, which multiplied our powers exponentially.

Humans were no longer as ordinary — or stable — as chimps. We weren't inherently destructive, or rotten to the core, but our special abilities kept growing, and many of these abilities were very useful to cultures that developed destructive tendencies.

Many, many thinkers have presented opinions on how we got to where we are today. At one end of the spectrum, we have the human exceptionalists, who believe that humans are divine, godlike, supreme. We are where we are today because of Sacred Destiny — and this is good! Indeed, no one can deny that we are really clever at making tools. At the same time, our tools are causing immense damage to the planet's ecosystems, and this is most embarrassing to the cult of human superiority. We're looking at a rapidly growing global catastrophe, and we're trying to enshrine it as a glorious success — a monument to human intelligence. This daffy belief requires the skillful application of highly advanced magical thinking.

At the other end of the spectrum are the misanthropes, who believe that humankind is fatally flawed, and therefore certainly doomed. It's not hard to formulate an argument that humans are flawed. It's less certain that our flaws are indeed terminal and incurable. Nobody knows the future. The game isn't over yet. It simply defies the imagination that a species so clever and intelligent could knowingly destroy itself by being stunningly unclever and foolish. But here we are, in the growing shadows (gulp!). Will growing eco-awareness finally reach a tipping point, and suddenly change the course of human history? Stay tuned.

John A. Livingston, a Canadian ecologist, took a position that was less arrogant than the exceptionalists, and more charitable than the misanthropes. Instead of *fatally flawed*, he saw us as *overspecialized* — because of our dependence on language and learning. Our success wasn't based on the slow process of physical evolution — developing stronger legs or sharper fangs. Instead, our success was based on acquiring and using accumulated knowledge, which enabled us to make quick improvements in response to new challenges. If you knew how to make and use a bow and arrow, then you didn't need sharp fangs or fast feet. This collection of knowledge enabled us to survive and thrive in almost any ecosystem or climate. Polar bears and parakeets can't do this.

Humans are social critters that live in groups, and we've learned a lot from each other, over time. We learned how to cooperate and share. We learned how to communicate in a highly sophisticated way. We learned how to make and use tools, in a manner more complex than other animals. We learned how to pass detailed accumulated knowledge from one generation to the next. We learned how to make and use fire. Because of this accelerated mental activity, our brains exploded in size.

Humans were much better able to plan and execute a successful hunt than a pack of wolves — and we knew it. We also developed the

What Is Sustainable

ability to kill and eat man-eating predators. We understood that we were extremely clever, and we sometimes got big-headed about it.

Overspecialization, according to Livingston, does not mean that we are certainly doomed. It means that, in an evolutionary sense, we are an unstable species, and therefore at high risk for extinction — like the overspecialized koala bear whose sole source of nourishment is eucalyptus leaves. Some scientists consider humans to be the most overspecialized species in the entire evolutionary journey.

Because of our high-powered minds, humans raised in unbalanced cultures are sort of like two-year-olds with a box of hand grenades — powerful, unpredictable, dangerous. Humans in healthy cultures do OK. There are still a few societies where people continue to live in a low impact mode that is very old. The quality of our cultural software has a major influence on the quality of our lives, and the health of our ecosystem.

During the Dust Bowl of the 1930s, "black blizzard" dust storms rapidly grew in size, as blowing dust scoured up more and more dust from the ground, and the whirling mass of airborne dirt rapidly grew until the dust storm was a huge black cloud, several thousand feet high, containing millions of tons of soil particles (search the web for "dust bowl photos").

I can visualize the history of big-brained humans in a similar manner. Ten thousand years ago, when our population was small, and widely dispersed, our collective mental powers did not create a high risk of large-scale mayhem. But, the more people there are, living in greater density, the more dangerous we become. The combined cleverness of a large herd becomes greater than the cleverness of its individual humans. This collective cleverness grows and grows, like a massive dust storm, until it becomes a terrifying disaster. Thus, we might think of industrial civilization as a massive and catastrophic *brainstorm*. Like dust storms, brainstorms are powerful, unpredictable, uncontrollable, and spectacularly destructive.

Nomadic foragers could never have made automobiles or computers, because manufacturing these things requires both immense collective cleverness, and large numbers of humans performing a wide variety specialized tasks in multiple locations in a highly coordinated manner. Foragers had no use for high tech stuff — it had no purpose in their sustainable way of life. If they wanted to go somewhere, they walked. If they wanted to jabber with someone, they talked to them.

It is no wonder that the early civilizations came into existence in places having an abundant supply of wild foods. The human herd ate well, and grew in numbers, which meant that the collective cleverness grew, too. Cities frequently brought together people from many different cultures. When specialists from different cultures communicated and shared, innovation thrived, often at the expense of tradition and stability. The emergence of reading and writing also fueled the exchange of information, the speed of innovation, and the destabilizing effects of change. Cities were hubs of political power, and their hunger for more power was generally constant and insatiable — they were the birthplaces of nations and empires.

Today, with high technology, we live in a global city in which innovation occurs at warp speed. Change that used to take centuries now occurs in a mouse click. The screaming roar of endless accelerating radical change in every sector of our society seems to have completely separated us from who we are, and where we came from. Our brainstorm is devouring life on Earth.

A gust of breeze blowing around a bit of dust is ordinary and insignificant. But, under the right conditions, a little breeze can turn into a huge dust storm. There is not an exact point at which a small sustainable society transforms into something that is seriously destructive — but the greater the population density, the greater the potential for destructive harm. Small is beautiful, as they say.

The good news here is that we are not dim-witted and helpless prisoners of our history. We are capable of mindful change. The fact that we are laying waste to the Earth's ecosystems right now does not eliminate the possibility of radical intelligent change in the future. In my own life, I am far less wasteful than I was 30 years ago — and this is the direct result of deliberate changes in the way that I think and live. In the world, environmental awareness has grown sharply in recent decades.

In theory, we have the power to break free from the trends of our history. Overspecialization in learning and thinking provides us with an unusual wild card. Our mighty brains give us the ability to foresee problems, to analyze our mistakes, to alter our patterns of thinking and behavior. In theory, we aren't doomed to continue repeating our mistakes. Self-destruction is not our only option. Will we play our wild card? I remain a dreamer and romantic, despite the huge odds. I see no harm in trying. We have nothing to lose.

The Curse of Progress

Most cultures perceive history to either be a saga of random events, or a tale of continuous decline. The modern concept of *progress* — that each generation is better than the one before it — is a bit over 200 years old. Marquis de Condorcet first popularized the notion in a 1795 book. Progress tells us that the Golden Age of humankind lies in the future, not the past. Every day, life is getting better.

Before long, the myth of progress became a core idea in Western civilization, and it took on a cult-like dynamic. It provided sacred justification for the brutal conquest of new colonies. We were actually helping the primitive people we were killing, because we were bringing them a better, more refined way of life — agriculture, literacy, Christianity, and civilization. Their children would thank us for our generous and benevolent gifts, according to the myth.

Here's the bottom line: the 200-year reign of the myth of progress has been the most tumultuous and destructive in the history of the planet. A person engaged in clear thinking could easily conclude that progress has been a great leap backward. Each unit of perceived benefit almost always seems to be linked to many units of indisputable drawbacks.

FORKS IN THE PATH

The Dark Juju of Stuff

Most early societies were nomadic. When the food supply around an encampment would diminish, the clan would pack their things and move to a fresh location in their region. An important benefit of nomadic living was that people generally didn't own more belongings than they could carry in two hands (except in the Arctic, where folks remained happy, despite owning kayaks and dogsleds). Possessions have been a major source of discord throughout the entire span of human history — even among people who owned almost nothing.

Common, readily accessible things like fruit, nuts, or berries caused little friction, because everyone was free to take what they wanted. But whenever anyone owned anything that's even slightly unique, it was likely that others would get jealous, envious, and covetous (similarly, dogs will fight over balls or bones). Therefore, in pursuit of group harmony, early societies kept their belongings in regular circulation by ritual gift giving — I'd keep something for a while, then I'd give it to you. You'd keep it for a while, and then pass it on to another. In this way, the potential for discord was reduced.

In North America, the fur trading industry blew this ancient balance apart. It offered manufactured products that the Indians found to be really useful (guns, knives, axes) or highly desirable (whiskey). The more furs you had, the more stuff you could get at the trading post. Old taboos against overhunting became obsolete. In many regions, beaver populations were brought to the brink of extinction.

Some groups shifted away from their traditional life and made fur trapping a primary activity — corn growers quit farming because trading reaped more benefits with less effort. The fur trade generated lots of friction between individuals, families, and tribes. The Beaver Wars raged, off and on, between 1629 and 1701. Many tribes were driven off their traditional homelands. Thousands died. Alcoholism became widespread. When the beavers became scarce, the trade ended, and it was nearly impossible to return to the old way of life.

My wild German ancestors suffered a similar experience. They allowed Roman traders to travel in their lands, because they provided unique and fascinating stuff. The Germans acquired axes, knives, and glittery stuff by trading cattle and slaves. This led to raiding, cattle rustling, inequalities of wealth, and the eventual breakdown of social harmony. The traders tried to offer wine, but the Germans refused it, because it made them soft and womanish.

Roman generals pacified German chieftains with generosity — land, glitter, lavish banquets, luxurious baths, education for their sons. Before long, the formerly fierce chiefs were wearing togas and speaking Latin. Their high status lifestyle was maintained by obeying the Romans, instead of protecting the people of their tribe. Throughout history, millions have submitted to a life of servitude in exchange for glitter and status. Countless consumers spend their lives doing unpleasant work in order to acquire cool stuff that they don't need.

Not all early societies were nomadic. In places having a plentiful supply of wild foods, people built permanent homes and lived sedentary lives. This commonly occurred in places with large salmon runs or an abundance of shellfish — like the Rhine River basin, the Thames River in England, coastal Japan, or the Pacific coast of North America. In the Middle East, people settled in regions having vast seas of wild wheat.

The sedentary way of life was more troublesome because people with permanent homes had the ability to acquire and store large quantities of material goods. In these villages, some people acquired more belongings, while others acquired less, and this created social friction.

Among some Pacific coast tribes, this resentment led to the creation of *potlatch* ceremonies, during which the rich would generously give away their excess belongings to the poor — or even destroy surplus stuff — in an attempt to balance out the distribution of wealth. Potlatch was often effective at reducing community tensions.

The Cherokees went even further than the potlatch rituals. Early every summer the Green Corn Ceremony was held to purify the tribe. This included giving away, burning, or otherwise destroying everything they owned. Every year, every family burned their house down. This was a culture that really discouraged the accumulation and hoarding of stuff. House-burning had the additional benefit of roasting the nests of fleas, including their eggs.

The Shoshone were generally peaceful nomadic foragers who owned very little stuff. When strangers passed through the neighborhood, the Shoshone simply vanished into the shrubbery and became invisible. They had no fields, villages, granaries, or livestock to protect.

On the other hand, when hungry strangers came to visit the corn-growing Cherokees, the Cherokees had to get defensive. Their corn couldn't run into the bushes and hide. Nor could their villages full of stuff. Their grain was their survival. If they lost it, they were doomed. Farmers didn't run and hide, they fought and died.

I have met highly educated, very intelligent people who believe that the desire to own, accumulate, and hoard material belongings is a core human trait or instinct. The anthropological evidence does not support this. *The desire to acquire and amass stuff is NOT a normal, natural, innate human trait.* Hoarding stuff as exclusive private property is primarily a characteristic of sedentary societies — a recent and unusual development in human history. The most extreme forms are found in contemporary consumer society, which is obsessed with hoarding. Shopping has become the purpose of life.

The dark power of personal belongings was so intense that one of the Bible's Ten Commandments specifically addressed it: thou shalt not get emotionally attached to other peoples' stuff. Jesus went even further on the subject. When a rich man asked him how he could experience heavenly paradise, Jesus told him to give away his stuff. Jesus was no fool.

In addition to stimulating conflict, materialism is also a primary contributor to the Earth Crisis, because so many consumers devote so much of their life energy to the acquisition and display of unnecessary conveniences, fashion accessories, and frivolous status trinkets — in a futile attempt to ease their inner pain.

Consumers measure the value of their existence by the quantity and quality of stuff that they have accumulated. It's an overwhelming obsession. They must always have more than they had last year, more than their parents had, and more than their neighbors have. They are cursed with a never-ending compulsion to "get ahead" — a painful itch that compels them to continuously hoard more stuff. They never experience contentment.

The Dawn of McAnimals

Humans originated in Africa and then gradually migrated to other continents. Around 10,000 BC, an episode of global warming brought the last Ice Age to an end. Huge glaciers melted, and this increased the water level in the seas and oceans.

This was a period of massive ecological change. Land that used to be buried under ice was now exposed to the sun, melted, softened, and turned green, enticing wildlife into formerly uninhabitable lands. Many regions of open tundra gradually turned into lush forests and prairies. The planet exploded with life.

In this changing reality, humans had to make adjustments to their modes for survival. Most cultures remained on the time-proven path of nomadic foraging. Some cultures began experimenting with new modes, which included the domestication of animals.

Humans have lived near thousands of species of birds and mammals for hundreds of thousands of years, but only a few dozen species of animals have allowed themselves to become domesticated. Thank goodness that most species have never surrendered their wildness! Imagine how dull the world would be if everything was passive and tame — cuddly slobbering Siberian tigers rolling on their backs for tummy rubs.

Domestication, of course, is a euphemism for enslavement — owning and controlling others. It represented a radical shift from the past. We robbed the freedom from some of our animal relatives, and then bred them into new forms that best suited our whims and needs. The new forms had smaller brains, and were less intelligent and aggressive, making them easier to control and exploit. See Paul Shepard's book, *The Others*, for a fascinating discussion of this.

The spirits of my wild ancestors talk to me in my dreams. They wish to go on record here. They do not approve of freedom stealing. It is not the path of balance and honor. In their eyes, domestication was a clear and tragic symptom of spiritual disease. It grieves them to observe how many of their descendants are now controlled and exploited by hard-hearted masters. They look forward to the happy day when all life is wild and free once more. Living beings are not meant to be property (or human resources).

Lame Deer was a Lakota medicine man, and he had strong views about wildness and freedom. He believed that wild buffalo had wisdom and power (and a sense of humor), but docile man-bred cattle were dullards. A sheep was pathetically weak. It would stand still while you cut its throat. He said that a wolf has great power, but white society had turned him into a freak — the pitiable lapdog, the most humanized animal of all. Lame Deer regularly enjoyed dog soup.

Dogs were the first domesticated animal. All varieties of dogs trace back to the wild gray wolf. In modern America, dogs are enjoying a peculiar level of popularity, compared to their past. Throughout history, most dogs were loud, dirty, foul-tempered scavengers who lived on the fringes of society, eating garbage, rats, feces, and corpses — and sometimes livestock, children, and other unlucky things.

In the Islamic world, the dog was seen as an unclean animal that drives away angels, annuls prayers, and limits their owner's benefits in paradise. Nor were dogs revered in the Hindu, Jewish, or Christian traditions. In Norse myth, the savage hound Garm guarded the gate to the land of Hel, the goddess of death. In Greece, Apollo's son Linos was eaten by dogs. The Aztecs raised small hairless dogs for meat. Dogs were food in many cultures, and still are.

The modern pet industry is extremely profitable, and massive advertising campaigns have effectively turned dogs into "fur children" — members of the family, who require premium food, expensive accessories, Christmas presents, limitless health care, and a cemetery burial with a carved granite headstone. In my town, some purebreds have been selling for a small fortune. The marketing and movie industries have been highly successful at turning dogs into status objects, fashion accessories, and fantasy projections. This is a stunning example of the power of marketing.

Like humans, the population of dogs is growing sharply. There may now be 500 million dogs in the world. A 100 pound American dog consumes twice as much food as a 100 pound human, and dogs are served a high-impact diet of processed industrial food. The ecological cost of dogs (and cats) is far from trivial, and this cost is very hard to justify in a crowded, beaten, and wheezing world. Our wild bird relatives do not speak highly about the millions of murderous cats who run loose around human settlements.

Millions of people acquire millions of pets and then dispose of them when they become inconvenient. US animal shelters kill six to seven million dogs and cats every year. Los Angeles alone sends 200 tons of dead dogs and cats to rendering plants every month, where they are converted into protein feed, and sold to fish farms, shrimp farms, and other outlets.

It's not just marketing that drives 39% of US households to own dogs, and 34% to own cats. Humans are animals, and we evolved in a world where we were surrounded by an abundance of other wild animals, and this constant exposure was phenomenally important to us on many levels. We have a profound hunger for contact with our wild animal relatives — in a wild landscape (zoos are horrid!). Despite how we live today, the normal and healthy mode for human existence is to celebrate our lives in a wild land, surrounded by wild animals — this is our ancient home, and the environment in which we thrive.

I have experienced this myself. In the Keweenaw, I spent nine years in the woods, and I had far more contact with wild animals than with domesticated animals, including humans. This was a great pleasure. There is nothing more noble and inspiring than a wild animal joyfully flourishing in a healthy wild land.

It's wonderful to go out at dawn and see a wild deer grazing beside her spotted fawn. It's wonderful walking on a moonlit summer's night, surrounded by a happy halo of fluttering bats, feasting on mosquitoes. It's fascinating to watch a gorgeous red fox chasing a racing swerving white hare across the frozen pond. It's amazing to come across a large black bear in the woods, dining on juicy yellow apples. It's delightful to listen to two owls hoot back and forth under a night sky throbbing with northern lights. It's thrilling to go to the woodpile and meet my good friend, the weasel, a wee creature with intense intelligence. The ravens were enthusiastic socialites, who generously made room for newcomers at a fresh banquet. Ravens are extremely smart, some of them can live to be 80 years old.

The coyote is a wild sacred animal of power who reminds us of what we once were. The neurotic, submissive dog is an infantile animal that reminds us of what we have been reduced to. Owning McAnimals is not an effective substitute for the life of wildness, freedom, and community that our injured spirits so desperately crave. There is no substitute.

After dogs, we domesticated herbivores (sheep, goats, cattle, pigs) and birds (ducks, geese, turkeys, chickens).

In the old days, hunting was a deeply meaningful spiritual activity. It required strength, mental clarity, paying acute attention, and utilizing a wealth of ancient skills. The age of domestication replaced the thrills and excitement of hunting with the monotonous non-stop boredom of herding — watching over docile grazing animals, day after day, year after year. It diminished both the herd and the herders.

Like dogs and humans, the population of domesticated livestock is also rising sharply — while the numbers of wild animals fall. Lame Deer said that there was no profit to be made from coyotes or wolves, so we drove them out to make room for profitable domesticates. In a sustainable future, coyotes and wolves will once again be honored and respected members of the family of life — free to live as they wish, in a world where varmint exterminators have gone extinct.

Personal Property

With domestication, folks quit seeing animals as our equals, teachers, and powerful sacred relatives. We began to believe that we were the masters of the world, and that everything else was below us. We began to think and behave as if we were gods. We began to see our enslaved animals as exclusive personal property — *my* chicken, *my* goat, *my* ass. Wild animals that raided the herd, chicken coop, garden, or granary came to be seen as diabolical pests that had no right to existence.

With the enslavement of animals, herders came to own many tons of walking meat, and this lead to a sharp increase in the dark juju of stuff. No one owned the huge, powerful, and ferocious wild aurochs, but the man who owned a passive domesticated cow claimed exclusive rights to its meat, milk, manure, and hide. Naturally, the man who owned a cow would be envied by the man who owned none. Because of this, the domestication of animals created exciting new career opportunities for humans — cattle rustlers, horse thieves, chicken snatchers, and so on.

With herding, the dark juju of stuff also escalated from the personal level to the tribal level. For example, the Navajos were originally nomadic foragers. After the Spaniards came, and the Navajos acquired domesticated horses, cattle, goats, and sheep, their traditional culture was thrown into chaos. Ownership conflicts arose, and this gave birth to a painful and bloody era of tribal warfare. Nomadic foragers rarely fought over wild animals, because there was little reason to.

Like many herding cultures, the Navajo protected their herds from predators, which led to the survival of more and more livestock. By the 1930s, the size of the Navajo herds was far in excess of the carrying capacity of the land. Large areas of grassland were stripped of vegetation and suffered from severe erosion — a permanent injury. My Welsh ancestors also had this experience, as have most herding cultures. This damage rarely happened when wild animals lived in a wild land.

Herding and raiding traditionally went hand-in-hand. When Native Americans acquired horses, they became horse thieves. The cattle herding Maasai of Kenya took great pleasure in cattle rustling, even if it required killing people. The same was true of the cattle herding Celts and Germanic tribes in the Roman era. They thought that a cow was far more valuable than the life of someone in a neighboring tribe. Because of frequent raiding and endless feuds, herding cultures tended to be warrior cultures, like the famous Mongols and Huns.

While hunter-gathers were nomadic, they wandered within a home region that had specific and ancient boundaries. These boundaries were generally honored by neighboring groups of people, who had their own specific home regions to hunt within. Occasional trespassing was tolerated, and not uncommon. But overall, boundaries were respected.

With the domestication of animals, land ownership became a growing source of conflict. Herders were control freaks, and the successful ones excelled at enforcing claims of ownership — by bloody force, if necessary. If the sheep of neighboring clans came for a visit and ate your grass, then your clan would be deprived of milk and meat. This trespassing could not be tolerated by people who had a growing belief in exclusive ownership rights — fetch the war paint, lads!

Ecological Impacts of Grazing

The domestication of animals has contributed significantly to ecological destruction over the centuries. There are a number of problems associated with herding:

- Overgrazing inhibits the regeneration of forests, as seedling trees get eaten for lunch. For example, sheep have inhibited the return of the forests in the United Kingdom, and goats have prevented the return of the forests of North Africa.

- Overgrazing exposes the soil and encourages erosion. For example, in China and Inner Mongolia, goats are converting large areas of fragile grassland into desert, in order to supply consumers with bargain-priced cashmere wool. The goats devour the grasses, roots and all.

- Prior to herding, few Africans suffered from sleeping sickness (trypanosomiasis), which was spread from wild antelope to humans via tsetse flies. But herders lived in close contact with animals, and the insects they attracted.

- Herders typically exterminate wild predator species in order to minimize the loss of domesticated livestock. Lions used to be common in much of Europe, and western Asia. California's mascot, the grizzly bear, is no longer found in the state.

- Livestock herds displace wild herbivores. For example, the western United States used to be home to vast herds of wild bison. These wild bison were deliberately exterminated to make room for vast herds of enslaved cows and sheep. Likewise, the *wisent*, the forest-dwelling bison, was once common across Europe, from Russia to southern England. Like the American bison, the wisent herds were exterminated by the arrival of civilization. By the 11th century, the wisent population was reduced to a few hundred. Today, several thousand survive, but the wisent is an endangered species.

Unlike dropping a lit match in a dry forest, the destruction caused by herding was generally not immediate and dramatic. It often took a few generations to ruin a countryside. Slow motion destruction is much harder to see. Today's cowgirls have no memory of what the land looked like when their grandmothers were young.

In recent decades, there have been major changes in the way we raise livestock and poultry. Our cows and chickens are now jampacked into Confined Animal Feeding Operations (CAFO), where they are fed high-powered processed feeds and antibiotics, whilst wading around, shoulder to shoulder, in deep shit. Chickens can live their entire lives without having spent a minute in the sunshine. There are dairy cows that don't know what grass tastes like. Manure is produced in such quantities that it is no longer a valuable fertilizer — it has literally become toxic waste.

The War Horse

In the old days hunting was not easy. Wild grazing animals were very fast, while humans on foot were very slow. For example, an American bison can run at speeds up to 35 miles per hour (56 km/h), and they do not graze placidly while hunters stalk them, they run. They have no inhibitions about charging, goring, and trampling hunters. But a big change occurred in about 4350 BC when humans in the Ukraine began to enslave horses.

Horse-mounted humans were amazingly speedy, and this enabled them to effectively chase and kill speedy game animals. The Spanish brought horses to the Americas. Some of these horses escaped and migrated northward, into the future US, where plains Indians readily became horsemen. This sparked a revolution — raiding increased, warfare increased, and many more buffalo were killed. With horses, a number of tribes quit living in villages as corn-growers, and returned to being nomadic hunters.

Peter Farb described the horse revolution in *Man's Rise to Civilization*. In 1800 the Bannock and Northern Shoshone acquired domesticated horses. Then, these horse-mounted Indians went hunting. Horse-mounted hunting created a temporary decades-long period of prosperity and population growth. By 1840 they had nearly exterminated the bison in the Great Basin.

In Eurasia, the domestication of the horse was a profound history-changing event. It revolutionized warfare, leading to oceans of spilled blood, and the creation of vast new empires. Armies could travel farther, make rapid surprise attacks, and quickly retreat. In open spaces, foot soldiers could not defeat mounted ones. The horse was a powerful battlefield weapon through World War I.

The unlucky residents of the Hungarian plains were not at all delighted to meet the vicious cavalry of the Mongol hordes, who raced across their lands, raping and pillaging. The Spaniards could not have easily conquered the Americas without horses (and smallpox). The horse became a symbol of the warrior and dominator. In many town squares you will see statues of mighty male heroes on horseback.

In his war god aspect, Odin rode into battle on his magical eight-legged horse Sleipnir. The Norse worshipped horses to such a degree that they considered horseflesh to be sacred food. When the Christians conquered them, the eating of horseflesh was banned, which is why it isn't for sale at our supermarkets today.

The horse also became a powerful tool in the war on soils. Because it moved faster than the ox, the horse enabled farmers to plow fields farther from the village, and to plow more acres per day. Loggers loved horses, too, because they could haul logs from locations that were previously impractical. So, fewer forests were safe from loggers, and many of these were converted to additional farmland. Increases in farm acreage led directly to increases in population, which led directly to increases in warfare and ecological destruction. Domesticated horses greatly increased the destructive potential of human societies.

Human Domestication

Once we learned how to domesticate livestock, we discovered that humans could also be domesticated, by using similar coercive techniques. The history of civilization is, after all, essentially a story about controlling and exploiting large herds of submissive humans. In our schoolbooks, the domestication of humans is euphemistically referred to as "the dawn of civilization" — the triumphant and glorious sunset of human freedom.

In the early days of colonial New England, the Indians drove the Puritans crazy. The Indians had real problems with bowing and submitting to the designated (white) Big Man. Indians only paid homage to individuals whom they chose to — and only because these chosen individuals were genuinely worthy of homage. The Indians thought for themselves, and made their own decisions. Can you imagine a world without rulers and slaves, without owners and property? Can you imagine being wild and free?

In our culture, the concept of *wild and free* has always had a superlative meaning — there is no higher state. Would you rather be a wild tiger or a tiger in a zoo cage? A wild timber wolf or a miniature poodle? Wild and free means unspoiled, untamed, unbridled, unlimited. It means pure, natural, healthy, good. It means strong, living life to the fullest, bursting with vitality. It means having incredible awareness, razor sharp senses, and a clear balanced whole mind. Wild and free is the ideal state, the zenith, the divine, the pinnacle of existence, the way things were meant to be — in other words, *normal*.

But in our culture, we rarely talk about wildness and freedom. We have a hard time even imagining what being wild and free would feel like. We have forgotten what it means. Something sacred inside us has withered. When a rabbit has been held captive long enough, it will remain in its cage even when the door is left open.

We are living a life that we were not born for. Every newborn child arrives in this world wild and unbroken. The process of taming a wild horse is called *breaking* it. My wild ancestors were not broken people, and they did not live in a broken world. The generations yet-to-be-born do not have to be broken McAnimals. Our cages are not locked. I pray that all humans everywhere will someday recover their wildness and freedom. The sooner, the better.

The First Farmers

Domestication of Grains

Many believe that farming began in a region known as the Fertile Crescent, in the vicinity of modern Turkey, Syria, Iran, and Iraq. A bit later farming independently emerged in parts of China, India, Mexico, Peru, and New Guinea. All of these regions were ecologically rich — wild foods were plentiful and easy to harvest with little work.

A key factor in the development of agriculture was the adaptation of foods that had an extended storage life, like corn (maize), wheat, rice, beans, or potatoes. Dried potatoes, grains, and legumes could be stored for years. These foods were rich in calories, and were a source of protein and other nutrients. They could also be fed to domesticated poultry and livestock, and converted into meat, milk, and eggs. So, they were a very useful form of food energy.

The Fertile Crescent was originally a forager's paradise. The wild grasses were plentiful and many of them produced edible seeds, including barley and three members of the wheat family — einkorn, emmer, and spelt. There were also large ancient forests, lots of fresh running water, and abundant herds of wild grazing animals. It must have been similar to the Garden of Eden.

Around 1980 a scientist named Dr. Jack Harlan visited the region, fetched a sickle, went to a stand of ripe wild einkorn, and commenced to forage. He discovered that in three weeks a family could gather a year's supply of grass seeds. Not bad! There were also obsidian deposits in the region, so folks could make super-sharp cutting tools used for harvesting.

At some point, the foragers in the Fertile Crescent realized that it was possible to gather and store an entire year's supply of wild grass seed. If they built granaries (to keep out the dirt, rain, rodents, bugs, and birds), then they could store the dried grass seeds, give up the nomadic life, and live in permanent homes in permanent communities. That was what they did. They became sedentary foragers.

For a while, sedentary foraging worked quite well. People had enough to eat, and the labor required to gather the food was far from oppressive. But over time, the gathering of wild grass seeds shifted to planting domesticated grains, a much more laborious process.

Experts remain far from unanimous in their explanation for the birth of agriculture. It very likely had something to do with growth. There were a limited number of locations where edible wild grass seeds grew in abundance. Eventually, we found them all and used them. If we failed to carefully limit our numbers, or if we over-exploited the wild grain, then the supply of seeds would eventually become inadequate. We had to choose between fewer mouths or more seeds. We began planting seeds in new locations, and when we ran out of flood plains, we started clearing forests.

At about the same time, the humans of the Fertile Crescent began domesticating herbivores — sheep and goats. By and by, an early form of organic agriculture was born. The emergence of plant and animal domestication represented a sharp turn away from the three million year Golden Age of sustainability.

Paul and Anne Ehrlich have written that the development of agriculture totally reorganized human society, as it increased the population. It freed a minority of people to become rulers and specialists. By and by, this led humans to temporarily become the dominant animal on Earth.

Old-fashioned low-tech organic agriculture was ecologically devastating. When farmers arrived, the healthy wild ecosystem promptly got decimated. Forests had to be killed to clear the land. "Weeds" had to be eliminated, so their seeds wouldn't spring to life in tilled fields. Wetlands had to be drained. Critters who ate livestock had to be systematically exterminated — the hawk, owl, wolf, fox, coyote, weasel, felines. Likewise, critters that ate grains, fruit, and veggies also had to go — the bunny, deer, raccoon, bear, birds, rodents. Following the arrival of farmers, the original wild ecosystem was largely erased, and replaced with an artificial human-controlled ecosystem that was way out of balance with the natural order. Man became the lord of the land, temporarily.

In the Keweenaw, my large garden did very well for the first two seasons. Then, the deer, coons, and hares decided to help themselves to the delicious food. It was sickening to discover that 30 feet of beets had been wiped out overnight. They kept coming back for more, leaving me with just the potatoes and onions. I remember sitting under a bush with a shotgun on a hot and humid day, furious at the wild garden-wreckers, while hundreds of buzzing gnats and mosquitoes dined on me. Wildlife is the farmer's enemy — and farmers are the enemy of wild ecosystems.

Nomadic Security

Nomadic foragers enjoyed high security, because they were conservative. They didn't regularly eat every type of edible plant in their territory. Generally, the less desirable food plant species were left untouched and only eaten during lean years. Some groups set aside portions of their land that were only used during lean years.

In his book *Man on Earth*, John Reader described how the !Kung bushmen of Botswana easily survived during the third year of an extreme drought in the 1960s. While 250,000 cattle and 180,000 farming people had died, the !Kung were providing their food with just 12 to 19 hours a week of effort. They dwelt in a desolate "wasteland" that no civilized people could survive in, and they lived well and joyfully. Conservative living is a beautiful thing, eh?

Nomadic foragers were not growth-obsessed people. They knew how many people their land could support, and they kept their numbers well below this limit. They maintained a huge safety cushion of food, for emergency use only. They used various methods of population control — herbal contraceptives, abortion, extended nursing, abstinence, infanticide. By deliberately choosing to live conservatively, they avoided war, starvation, ecological destruction, overpopulation, and the horrors of civilization.

It was an utterly brilliant strategy. Like the person with a billion dollars in the bank, nomadic foragers never worried about tomorrow. As long as they kept their numbers small, and their safety cushion large, the future was not a source of fear. Whenever their food supply ran low, they simply packed up their things and went where the eating was better. In extreme years, they ate their safety cushion. It's important to remember that these were real people with a long history — they are our wild ancestors, and their blood runs in our veins.

Sedentary Insecurity

The nomadic foragers enjoyed relative security because they lived conservatively. The sedentary farmers of the Fertile Crescent, on the other hand, took a very different approach. They built more durable shelters, and their welfare was largely dependent upon the performance of a few species in the nearby fields and pastures. If the season was bounteous, they thrived. If it was lean, they starved. It was a risky way of life — a huge gamble.

One day you could be looking at an awesome bumper crop ripening in the sunny warm valley. The next day could bring disastrous winds, a hailstorm, or fire. Your harvest could be destroyed by insect pests, birds, animals, diseases, tornados, floods, or raiders. Too much rain, or too little. Too cold, or too hot. An epidemic of rinderpest could suddenly eliminate the cattle, oxen, sheep, and wild grazers in the region.

In the years 1845-1847, one million Irish starved to death in the wake of a potato blight. In 1943, three million starved in Bengal, in the wake of a rice fungus. In 1970, corn blight raced across the US, during a wet summer, threatening to destroy our corn monoculture. Some southern states lost 50% of their crop. A food supply disaster was avoided by a shift in the weather.

In his book *Late Victorian Holocausts*, Mike Davis provides us with vivid and horrifying images of 19th century famines caused by El Niño droughts. There were two major drought periods that struck about 1876-1879 and 1896-1902. An estimated 30 to 61 million died in India, China, and Brazil. These three countries were hit hardest, but many other countries also suffered high mortality.

Farmers faced countless challenges. Seed harvesting had to be done at just the right time — not too soon, and not too late. Heavy rains at harvest time could ruin you. You could be lucky, and successfully harvest the bumper crop — but you were still vulnerable to rodents, fire, invasion, and thieves. The grain had to be kept protected and dry until the day you used it. For the sedentary farmer, total disaster could occur at any time. It could not be predicted. Often, it could not be prevented. His future was perpetually at risk and uncertain. When hard times hit, he was in big trouble.

More Is Better

Farming is an unreliable business. In many years, the harvest is less than average. To provide some insurance against periodic shortages, the smart farmer made every effort to produce food in excess, and then hoard the excess. Grain had a long shelf life, and what you didn't eat this year you could eat the next. So, for farmers the cardinal rule for survival was: *more is better*. This approach gave them a safety cushion of extra food. Unfortunately, it also stimulated population growth — in agricultural societies, more food very commonly leads to more mouths, because extra workers are an asset.

One serious problem that routinely confronted every farmer everywhere and every year was soil depletion. Each harvest took a toll on the health and fertility of the fields, which reduced future yields. In some places, primitive farmers could use a field for just one season. Richer soils allowed several seasons of use. Thus, primitive farmers had to move periodically, to avoid starvation. Later, farmers learned how to slow the rate of soil depletion by applying compost, manure, and other treatments.

Growing population and ongoing soil depletion drove farmers to regularly clear new fields, so they could plant their seeds in fertile soil. All farming communities had an ongoing need for fresh soils to destroy, but the land surrounding each village was finite. There were boundaries. All growth will find a limit. Planet Earth is finite, and so infinite growth is impossible.

Once village expansion had consumed and exhausted all unclaimed land, then the only path to continued production became the conquest of other lands. As the need for growth and conquest became almost mandatory, conquerors rose in importance. Warrior became a full time job, and war became a way of life.

Once the people in a region became completely addicted to agriculture, it became a vicious cycle. You couldn't stop growing. If your kingdom quit growing, then the growing kingdoms around you would grow right over you — so the choices were grow or die. Growth became a deeply rooted addiction and compulsion — the correct answer to any question was MORE! Long term survival depended on the creation and expansion of empires. It was a bloody solution to the never-ending problem of soil depletion.

Following World War II, Australia felt increasingly vulnerable to China. The Aussies worried that they were too weak to defend themselves from their growing neighbor. They actively encouraged immigration, spurred on by the notion of *populate or perish* — despite the obvious fact that their population already exceeded the carrying capacity of their mostly-desert nation.

Tragically, there was a real basis for the Australian's fears. The Tibetans, Mongolians, Taiwanese, and Koreans all know the fear of living in the shadows of a powerful and ever-growing giant. The Aborigines learned this lesson the hard way, as did the Native Americans, Indians, and Africans.

More is better is an idea that is deeply imbedded in our modern culture. They've done surveys, asking people of all different classes to specify the level of income at which they would feel comfortable, satisfied, and content. What they found was that no matter if people were homeless paupers or billionaires, the most common answer was "twice as much as I'm making now."

Driven by the perpetual hunger for more, agricultural society turned into an arms race. Whoever had the most arms (and legs) for fighting, breeding, and farming was going to come out on top. Peace became impossible, as did stability, and sustainability. Villages became towns, towns became cities, then industrial complexes, then megalopolises. Growth meant more taxes, bigger armies, and powerful kingdoms.

To a tragic and ridiculous degree, the concept of perpetual growth has been elevated to something close to a natural law — a sacred eternal truth that cannot be challenged. "Grow or die," say the liberals. "Grow or die," say the conservatives. "Grow or die," say the religious leaders, the educators, the business folks. Why? Why cling to an insane, self-destructive belief?

In a healthy world, more is not better. The key is simply to have *enough*, and be content with it — and "enough" is not much. Nomadic foragers were wealthy because their needs were small, everyone had everything they needed. Unlike modern consumers, they did not suffer from a persistent, insatiable need to acquire more and more stuff. Small and simple is the path with a healthy future, if neighboring cultures allow it.

Exploiting Domesticated Humans

Nervous and Irritable Crowds

As the centuries passed, the herd of domesticated humans kept multiplying, and the human-dominated regions expanded in area. This growth created a new and profound predicament. Evolution had not designed us to thrive in crowds, and we had no ancient cultural wisdom for responding to the stresses of overpopulation.

Nomadic foragers typically lived in clans of 10 to 30 people, because the hunting way of life prohibits living in large groups. Small numbers allowed the entire group to have regular daily contact and maintain close relationships. With larger groups, the quality of the interpersonal relationships diminished — there was not enough time to keep up with everyone. When this happened, society fragmented into subgroups, each having a different twist, and the harmony and cohesiveness of the society could shift toward friction and dissent.

For example, an office of ten people can work pretty smoothly together, and solve their problems without a lot of oversight and authority. An office of 200 people, on the other hand, cannot function without multiple full-time managers — the herd is too large for the individuals to work out their problems directly. The larger the herd, the more you need a firm system of rules and rulers — or the entire operation will quickly become unstable.

Humans thrived in small groups. Larger groups were unnatural and problematic. Humans got stressed and panicky when crowds of people surrounded them, especially when many of these people were strangers. Crowded situations resulted in increased levels of emotional instability, mental illness, and interpersonal conflict.

Added to the stress of overcrowding was the stress of the dark juju of stuff. With the shift to domestication, humans adopted a sedentary lifestyle, which enabled the acquisition of more personal possessions — houses, livestock, furniture, tools, etc. Some owned more, some owned less, and this inequality of wealth led to sharp social strains.

Shepherds and Sheep

To keep the human herds under control, leaders emerged. At the local level there were tribal chiefs who managed a village. These chiefdoms were overseen by regional warlords or kings. Kingdoms were overseen by mighty emperors, who ruled many regions. At each level of this hierarchy, leaders were more powerful. As their authority grew, their subjects surrendered more and more control over their own lives. By and by, this eventually resulted in a phenomenon called *civilization* — a hierarchical system for managing and exploiting high-density herds of domesticated humans.

Stored grain was an important form of wealth. The control of stored grain was likely the original form of political power. Rulers could determine who got fed, and how much. It was an excellent system for motivating people to be obedient and submissive. Control, coercion, and conquest were the keywords of domestication and civilization. It was about having power, using power, and increasing power. The system typically promoted ambitious, aggressive, and power-hungry people.

On a larger scale, it was the same story. Growth-driven cultures would devour cultures that were small, simple, and stable. Farmers always greatly outnumbered foragers, because they lived in higher density — so the foragers never had a chance. As the centuries passed, the farming cultures expanded while the foraging cultures shrank.

Sometimes the foragers were simply mowed down. Sometimes they were rounded up and sold as slaves. Sometimes they were coerced into assimilating the culture of the farmers. The conquerors often wooed conquered chieftains into voluntary subservience with objects

What Is Sustainable

of prestige and status, lavish feasts, cool clothes, intoxicants (rum, whiskey, opium, etc.). Easy credit from the trading posts quickly got many tribes hopelessly deep in debt, so they had to pay with their land. Children were sent to schools, where they were indoctrinated with the language and culture of the conquering oppressors.

All of our ancestors went through some variation of this process. Even today, in our literate and sophisticated society, many millions still voluntarily surrender their lives to the industrial system in exchange for cool stuff, easy credit, and powerful intoxicants. The pressure to surrender is intense and continuous.

The Anishinabe couldn't drive out the heavily-armed hordes of American oppressors. The Ainu couldn't drive out the Japanese oppressors. The civilized killing machine would take whatever it wanted, whenever it wanted to. The oppressors always won — in the *short run*. In the long run, the dance of perpetual growth and conquest would eventually self-destruct, because perpetual growth is impossible.

What does this mean for us today? We've been doing civilizations for maybe 10,000 years now, and we still can't figure out how to make them work (we aren't as clever as ants or bees). It's the same cycle — they grow and grow and grow, devour their resource base, and then collapse. They're still hotbeds for disease, highly polluting, destructive of soils, hell on forests, lethal to wildlife. We've never found a way to have a civilization that's stable, sustainable, secure, and happy — we've never come close to creating a healthy civilization — in fact, modern civilizations are even more destructive than the ancient ones.

Myths of Civilized Grandeur

Today, our myths tell us that the creation of civilization was a grand achievement. But it's interesting to take a second look at this. The oldest portions of the Judeo-Christian scriptures came from the days when the Chosen People were still nomadic herders. The Old Testament includes a number of prophets who conveyed God's angry demands to the Chosen People. One thing is clear — God hated civilization! Over and over again, God told His people to destroy the abominable civilizations that were spreading across the Promised Land. But the tide turned, farmers eventually overcame the herders, and the prophets of the Christian era have warned us that Armageddon is on its way.

The ancient Greeks saw human decline as a series of historic ages. In the ninth century BC, the farmer Hesiod wrote *Works and Days* in the dark age following the collapse of Mycenaean civilization. The ecosystem was devastated after centuries of abuse, and the population had plummeted to a mere 10% of its former peak.

In Hesiod's story, people of the long lost Golden Age lived like gods, pure in spirit, without toil or sorrow, in perpetual youth and continuous celebration. The Earth was healthy then, and provided food in great abundance.

The Golden Age was followed by the farmers of the matriarchal Silver Age, when men obeyed their mothers. This second generation was far less noble because they were simple, childish, and foolish. They did not give honor to the sacred, and were impure in mind and body.

The Silver Age was followed by a strong and terrible third generation, the Bronze Age. These people were hard of heart, violent, and warlike. They destroyed themselves and went to Hades.

The Bronze Age was followed by the farmers, explorers, and conquerors of the fourth generation — the Iron Age. Finally, these people were followed by a fifth generation that never rested from labor or sorrow. Disagreement was the norm — between father and son, guest and host, brother and brother. They were dishonorable, dishonest, and disrespectful — violent, foul-mouthed, and fascinated by every form of evil.

Thomas Jefferson, the former US president, had no affection for industrial civilization. Jefferson had visited the filthy cities of Europe, and they did not impress him in the slightest. Because of this unpleasant experience, he wanted the new United States to remain a nation of rustic farmers — with *zero* cities. He felt it was better to send American farm goods to Europe, in exchange for the manufactured goods produced in filthy stinking European cities. He thought that if Americans ever became urbanized, we would become as corrupt as Europe, and violently eat one another, as the Europeans were famous for.

If civilization was truly as wonderful as its boosters purport, then you'd think that the unfortunate uncivilized people would be pouring out of the woods, begging to join us in our wondrous and delightful existence. But this has never happened. Wild and free people have always detested and resisted civilization.

James Axtell was a scholar who studied how colonists and Native Americans interacted in early America. In *The European and the Indian*, he described a phenomenon that defies our sacred myths. He found that thousands of Europeans ended up living with the Indians, and they almost unanimously preferred the free and easy Indian way of life. Compared to the harsh rigidity of Puritan life, living with the natives was a lovely voyage of freedom.

These whites lived *voluntarily* with the Indians — by choice. When their Puritan relatives would come to drag them back to civilization, they nearly all resisted. They didn't want to go back. Connecticut actually passed laws banning whites from going off to live with the Indians — because their flight was draining the labor pool, and encouraging others to defect and flee. When civilized people got a taste of genuine freedom, they didn't want to return to the spiritually suffocating Puritan society.

On the other hand, Axtel found that Indians never voluntarily lived with the mean and racist Puritan Christians. When Indians were forced to live with the colonists, they made every effort to escape at the first opportunity. The Indians were proud and free people, not obedient servants. They had no desire to live like miserable civilized Europeans. Nor were the natives impressed with Christianity — while the

Puritans preached "love thy neighbor" at church, in their daily behavior this virtuous principle was often profoundly absent.

In the lands conquered by the colonial invaders, the Indians were in the way of the perpetual growth machine. The colonists insisted that the Indians adopt the European way of life, but most refused.

One tribe did obey the colonial orders: the Cherokees. They loved their ancestral home, and they desperately wanted to stay there. So, they chose to become civilized Christian farmers. They learned reading and writing. They wrote a constitution and created a democratic government. They built schools and churches and mills. They created a "respectable" society.

Then, the hateful racist white Christians passed a law, and confiscated their land, without compensation. The Cherokees were forcibly sent west, to Oklahoma. During the relocation process 4,000 of them were killed — one fifth of the people. This forced march was known as the Trail of Tears. Oppressors are not nice.

Heavy Metal

The Tasaday's Knives

John Nance wrote *The Gentle Tasaday*. This book describes the discovery of a small tribe of indigenous foragers in the forested mountains of Mindanao, in the Philippines. They were called the Tasaday. These people lived solely by gathering, and only had Stone Age tools.

Social workers made formal contact with the Tasaday in 1971, when loggers were moving into their region. At this time, the Tasaday were given steel knives as gifts. They found the knives to be amazing, and spent a lot of time chopping and slicing the surrounding trees and bushes.

The Tasaday talked endlessly about the power of their knives. They felt that before they had knives they were smaller beings, but that the knives had made them big — big arms, big heads, and big bodies.

Knives gave them a tremendous feeling of increased power, because they could now do many things that had previously been impossible.

We take knives for granted. We rarely consider how much more power a man with a knife wields over an empty-handed man. When steel knives were introduced to the Tasaday, it was a mind-blowing experience for them. It permanently altered their lives. The impact is hard for us to imagine.

Many experts now believe that the discovery of the Tasaday was a staged hoax. But the story of the power of metal tools has countless echoes throughout history. Indigenous people in many places resented being invaded by metal-making civilized people, but they found metal knives, axes, pots, and guns to be quite useful. Given a choice though, I suspect that most of them would prefer to return to their metal-free traditional way of life, prior to the arrival of the civilized folks — if only history had an undo button.

The Dark Power of Metal

In old Ireland, there were two classes of wee folk. The good ones, having *the bright knowledge*, were famous singers and dancers. They were called the *Tuatha Dé Danann*. These bright ones celebrated the living world, and never tried to exploit it with iron tools. In fact, they loathed and detested iron, as well as the people who made it and used it.

The dark wee folk, always busy at evil doings, were called the *Fomorians*. In the Old Norse stories, the miners, smiths, and jewelers were dark dwarves, who spent their lives beneath the ground, in cold deep caves where the sun never shined. Throughout these old tales iron is portrayed as dark and malevolent.

In many regions of Europe, the legends tell us that it was the unbearable cacophony of metal church bells that drove the wee folk from their ancient home. In the folklore of these regions, a good way to protect a civilized child from fairy mischief was to sew an iron pin into its clothes. Shamans often shape-shifted into wolves. The best way to kill a powerful werewolf was to drive an iron stake through its heart.

Metal had great power — it symbolically protected the civilized from the wild and the free.

The shaman and writer Martín Prechtel believed that technology took from the land, but gave nothing back. In an interview with Derrick Jensen, he once described the *spiritual economy* of the Mayan people. Every act of life had a spiritual price — nothing was free. If you picked a papaya, you had to leave an offering, to make the transaction fair and right — maybe a bit of tobacco. Items like steel knives had an incredibly expensive spiritual price, because of all the harm caused in their making — the mining, smelting, forging. Spiritually paying for a new knife required ceremony and offerings at every step of the fabrication. It was a long and complicated process. If the knife was not spiritually paid for, it would bring harm to the community.

In a number of languages, including Chinese, the words for *subjugation* and *violence* were connected to the use of metal. The Greek oracle at Delphi cursed the invention of iron.

Metal making had tremendous ecological and social costs. Axes, plows, and swords were weapons used in the war against the planet — destroying forests, soils, and primitive societies. Metal ores were finite resources, so no form of mining was sustainable. In fact, the mining industry was a huge source of toxic wastes. So was smelting, casting, and forging — it consumed vast quantities of energy, and emitted large quantities of pollution. Every step in the process was unsustainable.

The Iron People

In the early days of the Iron Age, iron production was limited by the primitive technology. One of the earliest iron daggers was found in Turkey, and dated to 2200 BC. But it wasn't until hundreds of years later that the Hittites made substantial improvements in the iron-making process. By 900 BC iron weapons were in common use throughout the Middle East. By 500 BC iron working was common in Germany and Scandinavia. The iron ore in Austria had carbon in it, which made for extra-sharp weapons.

There was an abundant supply of iron ore. When the quality problems were finally worked out of the production process, the iron industry began producing products in large quantities. Eventually these products became cheap, readily available, and popular.

Iron axes made logging much easier, and greatly increased the rate of cutting. More and more ancient forests were eliminated and replaced with croplands and pasture. With iron moldboard plows it became possible to turn heavy soils and thick prairie sods, which led to further expansion of cropland. Food production went up, and population exploded. Iron tools also spurred shipbuilding, international trade, and warfare.

Iron weapons made wars far more terrible, and led to a dark period of barbarism. When warriors with iron weapons went on the warpath, iron-free people were at a complete disadvantage. The spread of iron weapons was closely followed by the growth of military powers, including the Persians, Celts, Greeks, and Romans.

During the Iron Age the domesticated horse came into widespread use. Horse-mounted warriors with metal weapons were far more deadly than foot soldiers on open battlefields. Because of this huge advantage, kingdoms with armed cavalry found it easy to conquer simpler people.

It's easy to comprehend the power of metals when you compare the condition of Europe and America in 1491, the year before the Columbus invasion. America was largely covered with vast ancient forests, because it was essentially an ax-free Stone Age continent — the Iron Age had yet to arrive. Europe, on the other hand, had been devastated by logging — its ancient forests were largely gone by 1200.

Around 1000 AD, North America had a dangerously close call with the Iron Age. At this time, the Vikings attempted to settle on the Canadian coast at L'Anse aux Meadows in Newfoundland. They had brought a forge with them, so they could make iron tools. Luckily, for unknown reasons (fairy sabotage?) the settlement failed, and America's huge and healthy forests were spared for another 500 years.

After 1492 things changed. The beavers and buffalo were nearly eliminated. The cod and whales were decimated. The forests were mowed down. The indigenous peoples were exterminated. The Indians were deeply impressed by the destructive power and savagery of the Europeans, so they gave them a special name — the Iron People. I don't believe that it was a term of affection.

Sacred Forests

The Beauty of Forests

Living forests are a remarkable display of nature's harmony. The trees keep the land warmer in the winter and cooler in the summer, compared to open lands. A forest is like a giant sponge, absorbing huge amounts of moisture from rains and snows. This water is released slowly into lakes and rivers. In the good old days, there was far less fluctuation in the water levels in streams, between the wet season and the dry season — water-powered mills could be run year round on many streams.

There were a number of common side effects that often occurred when a forest was logged off (depending on local conditions):

- Many plant and animal species that enjoy living in forests cannot survive on stump land.

- The climate changed — rainfall diminished and the summers got hotter.

- The soils dried out and hardened. Water ran off hard soils far more quickly, increasing the possibility of flash floods.

- The land dried out. Springs, ponds, and creeks dried up. The water levels in lakes and rivers dropped.

- Soil erosion increased, which turned the rivers into muddy chocolate milk, which harmed the aquatic critters.

- Fish struggled to survive in streambeds that were no longer protected by forest cover. These exposed streams were warmer in the summer and cooler in the winter.

- In the mountains and hills, treeless slopes had a greater risk of avalanches, mudslides, and catastrophic erosion. The hillside soils became thinner and less fertile, in some cases eroding down to mineral subsoil, or even bare bedrock.

- Over time, sediment buildup made the rivers shallower and wider, and this increased the number and severity of floods. It choked the river mouths, filled in bays, and created sand bars.

For humans, no home was safer than an unbroken forest. Invaders were at a complete disadvantage — death could be hiding behind every bush in the misty dark shadows. The defenders knew every rock, tree, and stream — they could strike when it was advantageous, and vanish when it was not. The horse-mounted hordes of Mongols and Huns easily conquered the folks of the wide-open plains, but their empires stopped where the forests began.

But with the arrival of civilization and iron axes, invasive roads were hacked into the woodlands, and the safety of ancient forests was eliminated. Roads allowed the entry of loggers, law enforcers, dumpers, traders, invaders, missionaries, tax collectors, slavers, epidemics, and exotic species. Indeed, it is no coincidence that the words "road" and "raid" come from the same root.

When the European invaders first came to the Americas, those natives who lived near roads (like the Incas) were the first to be conquered. The road systems had been built by earlier conquerors to subdue and control the locals. When the new iron-armed horse-mounted Spanish conquerors arrived, these lovely roads made their successful invasion a piece of cake. Later, Red Cloud fought hard against the incursion of new roads through Lakota country. He called roads "thieves' highways."

Prior to the era of roads, the highways were made of water. Most of the world's large cities were founded near the shores of rivers, seas, and oceans. Boats allowed the movement of people and trade goods. The early boats were small and primitive. In the days of Julius Caesar, the fast-flowing Rhine River provided a formidable security barrier between the Germanic tribes and the Helvetti (Swiss). The Roman historian Tacitus wrote that the Suione tribe did not regularly carry weapons because they lived by the ocean. They enjoyed a life of peace and security because the boat technology at that time did not make it easy for raiders to arrive by water.

Then, around the 8th century, the Viking ship builders got really good. Using iron tools, they were able to transform dead trees into ships that could carry warriors and horses across the sea from Norway, and up the river to Paris. Being invaded by waterborne aliens was a mind-blowing experience for the Parisians. The Vikings inspired many cities to construct defensive walls. When the German cleric Adam of Bremen wrote in 1075, the Baltic Sea and the North Sea were thick with pirates. The only thing more dangerous than travelling by water was travelling overland, through the "frightful" dense forests of northern Europe.

Humbaba's Roar

The *Epic of Gilgamesh* is one of the world's oldest surviving stories, a product of one of the oldest civilizations, and written in one of the oldest alphabets — cuneiform characters, inscribed on twelve clay tablets. This story is from the birthplace of civilization, when the civilizations of the Fertile Crescent were still thriving. The *Epic of Gilgamesh* is the saga of King Gilgamesh, who built the city of Uruk near the Euphrates River, in what is now Iraq. He lived around 2700 BC — back when some of the vast ancient forests of the Middle East still survived (but not for much longer).

In a nutshell, King Gilgamesh killed Humbaba, the god of the forest, and then cut down every tree in the sacred forest. Not long after this, the whole region heard *Humbaba's roar* — the sound of the mighty rushing thundering flood. A primary side effect of deforestation is flooding, and these forest-whacking early farmers certainly understood this — if you mess with the forest, Humbaba is going to mess with you!

Gilgamesh's magnificent city of Uruk was completely abandoned by 700 A.D. Today it is a crude pile of brown rubble sitting amidst a desolate barren moonscape. But, to this very day, in the lands of fresh tree stumps, you can still hear the mighty Humbaba's roar. Some things never change.

Feeding the Fires of Industry

Civilization used wood for many purposes — buildings, fences, heating, wagons, tools, ships. Wood was also the main energy source for the industry of early civilizations. These industries made products like pottery, bricks, glass, and metal. Industrial complexes devoured massive amounts of wood, and this required massive amounts of logging, which resulted in massive amounts of ecological destruction. Logging cleared the land, allowing for the expansion of farming, which led to increased food production, which led to increased population, which led to increased deforestation. It was a vicious cycle.

By 1200 BC the countryside of Greece was nearly treeless, the soils were trashed, the population crashed, and people desperately emigrated to less spoiled regions. On Cyprus, the copper industry devoured four to five square miles of forest per year. The limekilns devoured four to five square miles per year. The pottery works devoured four to five square miles per year. So did residential heating and cooking. In Spain, 400 years of silver mining destroyed 7,000 square miles of forest. Industrial civilization devoured forests with head-spinning velocity. (John Perlin and Michael Williams have written two great books on logging history.)

The Greek philosopher Plato (427 to 347 BC) described the ecological catastrophe at Attica, which was typical throughout the region.

He said that the land was once rich and fat, forested and fertile. But now it resembled a sick man's skeleton. Lands that had once been arable had become barren. With the deep soil gone, the rains raced quickly to the sea. The springs and streams had dried up. The land had been laid to ruin. That's the story of civilization in a nutshell.

Italy was also once covered with healthy ancient forests — before the birth of Roman civilization. By the end of the third century, the woods were going fast, and Romans conquered other lands to seize their energy (timber) supply. They devoured the ancient forests of North Africa, and they devoured the ancient forests of Western Europe. Heating one Roman public bath required the consumption of 114 tons of wood per year. As the years passed, wood became harder and harder to acquire, and this energy shortage contributed to the collapse of the Roman Empire.

As the population grew, so did wood consumption. By the 1540s the iron industry in Sussex, England consumed 117,000 cords of wood per year. A large battleship required 50 acres of old-growth oaks. Some glass factories were semi-portable, and were periodically moved to new wooded locations. By 1600 most of the forests of England were depleted. London and coastal regions were forced to convert to coal burning.

When the Europeans crossed the Atlantic and conquered the Americas, it was the same story again. Forests were decimated to provide farmland, fencing, lumber, railroad ties, paper pulp, and fuel. A single sugar mill burned 90 acres of forest in a year — and there were *lots* of sugar mills.

It is important to comprehend the notion that industrial civilization began at least 2,000 years prior to the arrival of the coal-powered *Industrial Revolution* in 1800. Ancient Greece and ancient Rome were both industrial civilizations.

It is important to comprehend the notion that, wherever it was attempted, wood-fired industrial civilization promptly created massive environmental destruction — it was absolutely, completely, and ridiculously unsustainable. European ecosystems were already devastated by 1800, when the Industrial Revolution officially began. Thus, when we contemplate creating a healthy and sustainable way of life, far more is needed than simply turning back the clock 200 years.

My middle name is Adrian, which means a person from the region of the Adriatic Sea. I never understood the meaning of my middle name, because my family is not from this region. The Adriatic Sea was named after the Etruscan port town of Adria, a town located on an island several miles from the coast, offshore from the mouth of the Po River. Civilization ravaged the Po River watershed. The forests were decimated. With the trees gone, the soils washed into the sea, and destructive floods followed every big rain. Today, Adria is no longer a seaport on an island. It is a farm town, 14 miles from the sea. Adria's ancient Etruscan streets are buried under 15 feet of eroded soil.

The story of Adria has been repeated countless times. Much of the Tigris-Euphrates valley is now a wasteland. The vast cedar forests of Moses' Promised Land are now a withered desert. Greece, Rome, Turkey, North Africa — the entire Mediterranean basin has suffered devastating ecological injuries. So, my middle name reminds me of this story, and I never forget it.

Sustainable Forestry?

Today, loggers have vastly more powerful technology, enabling them to cut forests with amazing speed and efficiency. The keyword here is *efficiency*. The purpose of the modern technology is to reduce the cost of cutting and hauling logs — not to carefully remove trees whilst causing the slightest disturbance possible. Indeed, modern logging, like ancient logging, is a disturbing enterprise.

The modern logging process requires that roads be constructed to every corner of the cutting area. The roads often cross creeks, and go up and down hills. They are built on a minimal budget. The cutting process involves a lot of heavy equipment moving back and forth through the woods. The machine traffic pounds the dirt logging roads, loosening the hillside soils. The skidding of logs also tears open the soils. The combination of roads, traffic, and skidding provides ideal conditions for soil erosion whenever the wind blows, the rains fall, or the snow melts. After the cut, when new seedlings are growing, it is common to spray the land with herbicides, to eliminate competing vegetation.

With mechanized logging, it is virtually impossible to do the work without causing soil loss. Despite new techniques and strict new rules, substantial rates of soil loss are common. With each harvest, the health of the ecosystem is degraded, to a varying degree. Soil nutrients are removed when every log is hauled out of the forest. In many developing countries, the logging industry makes no effort whatsoever to work in a careful manner — their system is referred to as "rape and run."

What about old-fashioned low-tech logging? It's important to understand the effects of logging done by early civilizations. The Fertile Crescent and Mediterranean Basin were once generously covered with forests. Virtually all are gone today, and this region is now one of the most ecologically devastated places on Earth — some districts look like moonscapes. Mechanized industrial logging caused none of this devastation.

The Mother of All Forks

The Good Old Days

I recall one summer day, around 1957, at my grandmother's home in Detroit. In the sunroom at the rear of her house, I met my great-grandmother, Pauline. We shook hands, spoke a few words, and she gave me a dollar bill.

When Pauline Schmögner was born in 1866 there were 1.4 billion people in the world. When she died in 1958 there were 2.9 billion. The human herd had more than doubled in her 91-year journey.

When I was born in 1952 there were 2.6 billion people in the world. Today there are seven-point-something billion people. The human herd is quite likely to triple in my lifetime. In Pauline's 94 years the population grew by 1.5 billion. In my 59 years the population has grown by more than 4.3 billion. This rapid increase in population is anything but normal or good.

When I was born in 1952 there were no personal computers, DVDs, satellites, space shuttles, personal music players, disposable diapers, cell phones, jet skis, credit cards, shopping malls, microwave ovens, birth control pills, or fast food restaurants.

When my father Richard A. Reese was born in Detroit in 1913 there were no herbicides, antibiotics, plastics, microwave ovens, air conditioners, chainsaws, transistors, televisions, radios, nuclear reactors, atomic bombs, missiles, jet planes, semis, or tape recorders.

When my grandfather Richard Reese was born in Shawnee, Ohio in 1885 there were no airplanes, automobiles, refrigerators, aluminum products, or bulldozers. Flocks of passenger pigeons, containing millions and millions of flying birds, would darken the skies for days.

When my great-grandfather Richard E. Rees was born in Cwmbelan, Wales in 1843 there were no oil wells, metal boats, tractors, telephones, electric lights, sewing machines, toilet paper, repeating rifles, or dynamite. The Great Plains were covered by millions of buffalo. The Keweenaw was a pristine wilderness. Lake Superior was pollution free and filled with fish. I have spoken with older relatives who met old Richard when they were young. My, but how things have changed!

When my great-great grandfather Edward Rees was born in Wales in 1818 there were no railroads, cameras, bicycles, wooden matches, or telegraph systems. New York, Boston, and Washington were small towns surrounded by farms and forests. Detroit and Chicago were trading posts in the wilderness.

When my great-great-great grandfather Edward Rees was born in 1798 there were 900 million people on Earth. When he died in 1851 there were 1.3 billion. When he was born in 1798, Los Angeles had

300 residents, and the world population was growing by four million per year. When I was born in 1952 world population was growing by 45 million per year. In 2011 it was growing by 76 million per year. If a catastrophe struck today and killed five out of every six people on Earth, there would still be a bigger population than in Edward Rees' childhood.

Modern folks have a comforting belief about "the good old days," when the world was stable, people were kind, children were well behaved, government was honest, and life was simple and good. The good old days are generally seen as being within the realm of living memory — "when my mother was young..." or "when my grandparents came here..." If we could just turn the clock back 50 or 100 years, then life would be wonderful.

The spirits of my wild ancestors believe that the good old days in Europe ended several thousand years ago. They ended when the domesticators arrived, and our culture fell out of balance with nature.

A Brief Bubble of "Prosperity"

Clive Ponting wrote *A New Green History of the World*. It presented an ecological history of the planet, with emphasis on human activities and their environmental effects. The book was an immense act of scholarship, jam-packed with head-spinning quantities of facts and statistics.

Ponting was a history professor. Rather than studying events with a microscope, he sought the big picture instead. What were the trends that got us to where we are today, and where will current trends take us tomorrow? From his mountaintop view, with a 10,000 year perspective, one thing that was very clear to him was that the last 200 years represented a sharp break from the long-term trends in the history of civilized people. He realized that prior to 1800 *almost everyone, everywhere in the civilized world, lived on the edge of starvation.*

During the last two centuries, there has been an explosion of prosperity, fueled by an explosion in the consumption of non-renewable resources. This abnormal period will not survive a third century. There is not enough oil. There is not enough water. There is not enough topsoil. There is not enough everything, except for people, of which there are far too many.

The crash of 1929 was the result of an economic bubble, an endlessly inflating whirlwind of borrowed imaginary money. The crash of 2008 was caused by another storm of pathological borrowing. It is easy to see the history of civilization in the same manner — a swelling bubble of resources recklessly borrowed from other people, other species, other ecosystems, and unborn generations — a ponzi scheme. Nothing can grow without limit. All bubbles pop.

During the last 200 years, the lower classes have been eating a bit better, but they have not experienced much of the soaring prosperity. Only a very small minority has lived in an affluent manner (i.e., literate, reliable electricity, clean water, adequate nourishment, access to health care, etc.). The gap between the affluent and the poor is huge and growing. The prosperity explosion has been especially intense in the last 50 years. Most of us living in the affluent world are completely out of touch with how the average human — the majority of humankind — lives these days. We are isolated in a bubble of fantasies and illusions. This isolation will not last.

Looking toward the future, Ponting does not predict a blissful utopia. He concludes that are heading for a crossroads. One thing is obvious: we are moving into an era of shortages.

One more thought. For much of the agricultural era, the goal in life was mere survival — surviving to the next harvest, surviving beyond the next plague, crop failure, or invasion. Life was short and backbreaking. Today, in industrial societies, the purpose of life is to consume as much stuff and entertainment as possible.

Industrial civilization has provided us with many, many things — some useful, some harmful — but it provided us with *nothing* that we actually needed, nothing necessary for a satisfying and meaningful life, nothing that we wouldn't have been better off without. In retrospect, the colossal soul-killing, planet-killing costs of progress make the benefits tragically absurd.

Most of us spend our lives doing work that we don't enjoy. Then we throw away our excess wages on trendy doo-dads that we have no real need for. To a significant degree, our lives are devoted to consumption. The more money we make, the more we consume, and the more trash we discard. But owning a McMansion, three cars, and a huge flat screen TV does not bring us inner peace.

What was the purpose of our wild ancestors' lives? To live like human beings? To celebrate the perfection of creation? Perhaps they enjoyed an elegant and sustainable form of prosperity. Acquiring food was not time-consuming or backbreaking. Everyone possessed the necessities for sustainable survival. They had everything they needed, and no more. Nobody had too much food, stuff, wealth, or power.

Perfect Storm

The 200 year prosperity bubble has been something like an avalanche, one thing setting off another, a chain reaction. Let's take a quick look at what happened.

The Age of Exploration led to the European colonization of many regions — the Americas, Australia, Africa, and much of Asia. This led to a vast expansion of world cropland, much of it at the expense of forests and indigenous cultures. Rich European countries were no longer limited by the need to be self-sufficient in food. They could export manufactured goods in exchange for imported food from the colonies. They could grow like there's no tomorrow — and they did.

This spurred rapid industrial growth in Europe. Hungry peasants were lured into the cities where they could become hungry factory workers. There was expansion of coal-mining, iron-mining, steam power, railroads, steam ships, and lumber making.

At the same time, the agricultural productivity of the world grew explosively. The amount of food that we could produce every year skyrocketed, for many reasons:

- Grazing land has expanded 680% in the last 300 years.

- Cropland area has expanded 560% in the last 300 years.

- The water supply for irrigation was increased by drilled wells, motorized pumps, and the construction of many dams. Irrigated land is highly productive, and the area of irrigated cropland expanded sharply — 700% in the 20th century, and 3,400% in the last 200 years.

- Guano and mineral fertilizers boosted soil fertility in the 19th century, followed by high-potency synthetic fertilizers in the 20th century. In the last 60 years, fertilizer production has grown explosively.

- Herbicides, pesticides, and fungicides reduced crop losses. Their use has grown explosively.

- Farm machinery reduced labor costs. Tractors gave each farmer the ability to farm much more land. Powerful machinery enabled the plowing of prairie grasslands, and this expanded cropland area. In the old days, 30% of farmland was devoted to pasture to feed work animals, like horses and oxen. This land was now freed up for producing cash crops or livestock.

- Cheap petroleum in the 20th century accelerated mechanization, and the production and distribution of fertilizers and farm chemicals.

- Plant breeders provided a number of breakthroughs that enabled farmers to produce more food per acre. The development of extremely productive hybrid corn varieties began in the 1930s. This was followed by the Green Revolution, which created more productive varieties of corn, wheat, and rice.

- Farmers began planting highly productive food crops imported from other lands, like corn and potatoes from the Americas. Corn and potatoes generated much higher yields per acre. The supply of food grew sharply.

- Domesticated animals were imported from other lands. Cattle, sheep, hogs, and chickens were sent to the Americas, where they increased the food supply and produced manure for fertilizer.

- Improved transportation systems enabled farmers in remote regions to send their food to market, even to distant continents. Needed supplies and equipment became more available and affordable. Fewer people perished during droughts, because it became easier to deliver relief food.

- Agriculture changed more in the last 200 years than in all of the preceding centuries combined.

Until 200 years ago, population growth was fairly slow, due to malnutrition, famine, disease, and warfare. The birth rate was very high, and the death rate was very high. Since then, the death rate has been sharply reduced by better sanitation, nutrition, and medical technology — notably, effective vaccines and antibiotics. Unfortunately, the birth rate has not been reduced to a similar degree. Consequently, population has increased more than 600% since 1750.

Population pressure is a major cause of social conflict. The population explosion would have been far worse if the 20th century had not been a time in which stunning numbers of people died by atrocity — World War I (15 million), Russian Civil War (9 million), Stalin's regime (20 million), World War II (55 million), Mao Zedong's regime (40 million). These atrocity victims did not engage in further reproductive activities, naturally.

In his book *Throwing Fire*, Alfred Crosby provides a fascinating discussion on the evolution of weaponry. Frequent warfare demanded that industrial societies devote full attention to military strategies and technology. Innovative societies survived, while societies that slipped into a state of military inferiority became helpless sitting ducks. In a competition that never ended, society's brightest minds worked fiendishly to improve military effectiveness. The last 200 years has been a time of astonishing advances in military technology — machine guns, long-range artillery, airplanes, tanks, submarines, nuclear weapons, and on and on. We now have weapons capable of bringing human existence to an end.

The shift to coal as an energy source greatly increased industrial growth. Water power required that factories be built by fast-moving streams. Wood power required that factories be built by forests. Coal powered factories could be built in many more locations. Coal-powered trains and ships led to radical changes.

The shift to petroleum increased industrial growth even more. It spurred agricultural productivity, and it revolutionized transportation systems. It revolutionized everything. Industrial civilization shifted into fast forward. Today, our global way of life is totally dependent upon the existence of an unlimited and never-ending supply of cheap energy.

The modern consumer way of life bears almost no resemblance to how people lived 200 years ago. This is not meant to imply that civilized people lived sustainably in 1800. They did not. Their way of life bore almost no resemblance to the manner in which our wild ancestors once lived — in relative harmony with nature. Today, however, the degree to which we live unsustainably is breaking all records in the history of humankind. Every "improvement" has a cost. Most problems are the result of solutions. There is no free lunch. We have run up a huge bill.

The End of the Era of Cheap Energy

In May 1995, I saw a strange Mobil Oil advertisement in Newsweek magazine. The ad stated that, at the current rate of consumption, the world oil supply would last for another 43 years. I instantly got suspicious. Why was Mobil spending good money to tell us that there was no problem with the oil supply? Was there a problem with the oil supply? Yes, it turned out that there was.

I started asking questions, and a year later I received a fat package of articles from Walter Youngquist, a retired petroleum geologist and professor, who was busy working on a very important book: *GeoDestinies*. I got my baptism into the world of Peak Oil early, when the story was still on the lunatic fringe, and largely unknown. It was mind-blowing. I was going to witness big change in my own lifetime — a notion that had not occurred to me before.

Natural gas, petroleum, and coal are finite non-renewable resources, and we are consuming them at record-setting rates. There is currently much controversy with regard to the future availability of these fuels. Did world oil production peak in 2005, or is it still a few years away? Will natural gas production peak in 10 or 20 years, or will the new shale fracking technology significantly delay this peak? Will massive new deposits of oil and gas be found beneath the melting Arctic ice? Can coal last another 200 years, or will production peak in a mere 10 or 20 years, as proposed by Richard Heinberg? Nobody knows. The correct answers are certain to emerge with time, in retrospect.

Following the peak of production comes an era of declining production, which eventually leads to the end of production. Today, we are living very close to the historic peak of worldwide energy consumption, in all of its forms. Future generations will never again have the ability to live as wastefully as we do. Beyond the controversies over peak production dates, there are several important notions to consider:

- From the perspective of climate change (and for many other reasons), we should cease the use of these fuels today, and forever. Climate change has the potential to generate unpredictable catastrophes.

- On a rational planet, we would cease using nuclear energy today, and forever. We still haven't figured out a safe way to store the mega-toxic wastes for 100,000 years. We have never made a highly-complex device that was defect-free and totally foolproof. We have yet to evolve to the point where humans no longer make stupid mistakes — and boo-boos at reactors sometimes become first class history-making events.

- Alternative sources of energy will never come anywhere close to replacing fossil energy.

- The era of ridiculously cheap energy is absolutely over. We can have faith that the costs will continue to rise, but we cannot have faith that the supply will remain adequate or available. There will be more energy wars.

- One of the worst things that could happen to the planet would be the discovery of huge new deposits of fossil fuel, providing us with the ability to temporarily feed ten billion, and keep the global economy on extended life support.

- We are getting closer and closer to hitting bottom, finally. We are approaching a crucial turning point in the history of civilization — not because of reason and good judgment, but because we're getting close to the limits of growth. Before long, we'll have nowhere to go but up.

The end of the era of cheap and abundant energy is going to slow the production and distribution of everything — food, steel, concrete, chemicals, paper, electronics, clothing — everything! It means that we might actually be approaching the peak of human-caused ecological destruction (barring a major nuclear war). A massive population using powerful machines fueled by cheap energy can destroy the planet at a ferocious rate. But, a much smaller herd, using muscle-powered tools, is going to destroy the future at a much slower rate. Genuine progress at last!

The Return to Muscle-Powered Agriculture

With the conclusion of the era of cheap and abundant energy, our food production system is going to be forced to make huge changes, within several decades or less.

- The rising cost of owning and operating farm machinery will necessarily result in a return to muscle-powered agriculture.

- The return to muscle-powered agriculture will require that large areas of land be set aside to produce food for draft animals, like horses and oxen, if they are used. Each horse requires about five acres of good pasture.

- The rising cost and reduced availability of potent chemical fertilizers will diminish their use, which will diminish crop yields.

- The rising costs of herbicides, fungicides, and pesticides will reduce their availability and use. Yields will be reduced.

- The soils on many farms are in poor condition after decades of chemical use — nutrients have been depleted, and the complex microbiology of healthy living topsoil has been severely damaged. So, when powerful chemical fertilizers are no longer applied, yields will fall. It may take years for soil health to recover, and continued tilling will slow or prevent the healing.

- Cuba was forced to return to muscle-powered organic agriculture when the Soviet Union collapsed, and their ability to import cheap subsidized oil suddenly came to an end. This led to a huge decline in productivity. In the old days, they imported Soviet oil, but today they import 60% to 80% of their food.

What Is Sustainable

- Manure (and human poop) will have to be gathered and spread by muscle power. These organic fertilizers are far less potent than synthetic fertilizers, and their nutrients are less readily available for intake by the plants.

- A return to fallowing may be needed in some places, to allow depleted soils to partially recover, or to help dry soils replenish lost moisture. Fallowing takes land out of production, which reduces crop yields. Modern farming has replaced fallowing with the more effective practice of planting green manure crops (like clover) to rebuild the soil during the off-season. Because it works best in tilled soil, green manuring will require more labor than fallowing in the muscle-powered era.

- Cultivation, weeding, harvesting, threshing, hauling, and haymaking will have to be done with muscle-power.

- The rising cost of operating irrigation systems is going to reduce the area of irrigated land, which will drive down yields. Also, extensive water mining has depleted the supply of groundwater in a number of regions — permanently, in some cases.

- The high yields from hybrid seeds will diminish as fertilizer, chemical, and irrigation inputs decline. In reduced-input farming, old-fashioned non-hybrid seeds produce better results. Farmers will return to seed saving.

- Industrial mega-farms are too large for an easy transition to muscle-powered agriculture. So the size of farms will shrink, and their number will increase. Or, mega-farms will have to be worked by small armies of farm laborers.

- Many more farm workers will be needed. The current number of farmers in America is small, and their average age is roughly 55. Almost none of them have experience in muscle-powered farming. They don't have the animals, the harnesses, the wagons, the plows, and so on.

- There is going to be a huge surplus of workers in urban areas, and a huge shortage in farming regions. Many will be forced to migrate — and few of the migrants are going to be strong, healthy, and experienced at farm work. Imagine moving the idle millions in Los Angeles to the farm country of Nebraska — programmers, MBAs, movie stars, gang bangers.

- To a large degree, urban and suburban living will end. The return to muscle-powered agriculture means that farm labor will be performed by animals, not oil. Traditionally, muscle-powered agriculture required the labor of 90% to 95% of the population.

- Using muscle-powered agriculture to grow grain is the most challenging option, because it is extremely labor intense. More practical options include producing fruit, nuts, vegetables, meat, eggs, and dairy products.

The bottom line here is that big changes are coming. There is going to be much less food produced, and it is no longer going to be cheap and abundant. It will no longer be possible to rapidly ship food over long distances. Americans will no longer have the luxury of wasting half of the food they produce. Food will no longer be sold in bottles, jars, cans, boxes, plastic bags, or freezer packages. Our way of life is going to be profoundly different in the coming decades.

On the bright side, we're going to be returning to organic agriculture again. On the downside, it will continue to be extremely challenging to grow organic grains, tubers, and vegetables in a genuinely sustainable manner, for reasons discussed in the following two chapters.

What Is Sustainable

So, hopefully, muscle-powered organic agriculture will provide a *brief* transition period to a new and better era of sustainable living.

The Last Great Thrash

Our way of life has no long term future. The status quo is a dead end road. Awareness of the Earth Crisis has been growing in recent years, and a growing number of people have been starting to question our way of life. This makes the leaders of multinational corporations very nervous. Good consumers never ask questions, they shop till they drop.

In response to shifting public perceptions, a stunning transformation has occurred — every corporation has assumed the appearance of being eco-conscious, and everything they do now is "green" and "sustainable." Pesticides have become Earth Friendly, nuclear power has become safe, and filthy coal has turned into Clean Coal. It defies the imagination! It's almost too good to be true! We can solve the Earth Crisis by shopping our way out of it — hooray!

Solar panels, windmills, electric vehicles, LED lights, energy-saving appliances, nuclear energy, high speed railways, smart grids, smart cars, smart growth — none of this is harmless, and none of it is necessary for a healthy and satisfying life. Turning on a compact fluorescent light is an act of terror. Coal is violently ripped out of Mother Earth's breasts. Mercury and acids spew from the power plant's smokestack.

This "green," "sustainable," "eco-smart" era amounts to little more than the last great thrash of industrial civilization — its one last spasm of planetary destruction before the end of cheap energy pulls the plug on the global economy.

Wind turbines and solar panels are not made by forest elves in cute green suits. There are mines, chemicals, factories, transportation systems, fossil fuels. I keep thinking about the Mohawk nation of Akwesasne on the St. Lawrence River in New York. They live downstream from a number of industrial sites, including ALCOA and GM.

The fish are contaminated with methyl mercury and PCBs. The breast milk of Mohawk women contains high levels of PCBs and mirex.

Who is going to get harmed by the last great thrash? Whose babies are going to suffer defects from the toxic emissions generated by manufacturing awesome new electric cars? Whose wells will get poisoned? Whose rivers will become lifeless? Whose communities will become cancer hotspots? Who will die from the pollution generated by mining the minerals needed for making photovoltaic panels or green laptops?

The venerable British deep ecology philosopher, Reverend Henry O'Mad, sums it up like this: "In places that we never even heard of, young children, the hope and pride and joy of their parents, play games with drums of toxic waste. Everything we buy is a bomb, and these bombs kill innocent people indiscriminately. Violence and death spews from every orifice of our culture."

What we are doing is clinging desperately to our traditional way of life. We're working feverishly to maintain perpetual growth on a finite planet. Our sacred scientists are trying to invent new silver bullet solutions, in an attempt to keep industrial civilization alive and growing forever. We are invincible. The good old days of catastrophic growth and waste will last forever. We'll grow and grow and grow to the end of eternity. The Technology Fairy will save us. We will rule the entire universe, so help us God!

The last great thrash is simply a tragic and futile attempt to delay our culture's inevitable death. We are working like mad to keep a terminally ill patient alive for as long as possible, at any cost.

Alternative energy is a distraction, a false hope. Humankind has existed for maybe three million years. During our long period of relative sustainability, we relied completely on a free and non-polluting energy source — the sun. For three million years, the energy from the blazing sun has dependably fed the Earth's ecosystems, including us. Today, the sun continues to shine. There is no shortage of sustainable energy, but there is a serious shortage of clear thinking.

What Is Sustainable

Here's the bottom line: it's too late for a smooth, intelligent, carefully planned, and painless transition to a sustainable future. But it's never too late to wake up and get real. It's never too late to learn and grow. It's never too late to behave more intelligently. It's never too late to nurture a sense of reverence and respect for the living Earth, and the generations yet-to-be-born of every species.

Currently, we don't have the option of immediately returning to a life of nomadic foraging, or any other mode of sustainable living, because there are far too many of us, and the ecosystem is too degraded. So, in the coming years, much of our food will still have to come from farms. If we were rational, in combination with rapid population reduction, we would vigorously pursue less destructive modes of farming, starting immediately. How? The following two chapters present a basic introduction to the secrets and mysteries of the world of food.

CHALLENGES FOR AGRICULTURE

Consumer society creates countless products for sale. Almost all of these products are unnecessary for a healthy, meaningful, and enjoyable life — they are purely frivolous, and making them, using them, and discarding them causes much environmental harm. As we move beyond the era of cheap energy, the mass production of unnecessary products will eventually go extinct.

One genuine necessity is food. Global food production rose sharply in the last 200 years, but this growth is unlikely to continue through the 21st century. In fact, it's almost certain to decline, for a number of reasons. Meanwhile, population is projected to continue ballooning: to eight billion, then nine billion, then ten. Hello? I predict that this is going to be a memorable century.

The goal of this chapter is to examine a number of important challenges that confront the stability of agriculture. The path before us is one of big change and turbulence, and it is useful to understand the ways in which we are most vulnerable. A significant and embarrassing theme in history is one of repeated mistakes. Can we finally escape from this trap? My dream is that we can learn from our mistakes, and choose a different path — one that works.

Magical Topsoil

Topsoil is big magic. Each handful of healthy topsoil is alive with billions of living things: fungi, bacteria, algae, protozoa, grubs, insects, worms, and others. They work together to promote plant growth by decomposing dead material, converting nutrients into a form that plants can use, aerating the soil, inhibiting the growth of pathogens, retaining moisture, and creating new topsoil. Most of this community of beneficial organisms lives in the top two inches (5 cm) of the topsoil.

Simply, topsoil makes possible the existence of plants, and the animals that eat them. In his book *New Roots for Agriculture*, Wes Jackson reverently and affectionately refers to topsoil as the Earth's placenta. It is a thin living membrane where sunlight, air, water, and nutrients provide sustenance for life.

Agriculture is harmful to the placenta of topsoil — almost half of it is gone now. Agriculture has been destroying the placenta for thousands of years, and this destruction was greatly accelerated in the 20th century. There are entire regions of former cropland where farming is no longer possible. Many of the great cities of the Fertile Crescent are now abandoned piles of rubble surrounded by barren moonscapes. There are no alternative soils to switch to when the original soil is destroyed.

When the soil is turned over by shovel or plow, the living soil community is disrupted and damaged. Toxic chemicals used in modern farming to kill pests and pathogens also kill the beneficial organisms of the soil. When the life of the soil is disrupted, conditions are no longer ideal for plant life. Importantly, the process of creating new topsoil is slowed or stopped. Over the centuries, farmers have partially compensated for this destruction by adding nutrient supplements, like manure or compost. In modern industrial agriculture, nutrients are provided by infusions of chemical fertilizers. Wes Jackson says that the patient is being kept alive by artificial life support — chemotherapy.

When the topsoil is washed away or blown away, the placenta gets thinner. Sometimes erosion completely eliminates the placenta, leaving nothing behind but infertile subsoil or bare rock. American agriculture has damaged the placenta more rapidly than anywhere else in history. Iowa had an amazingly deep layer of topsoil, but half of it has been lost in just 150 years. Some churchyards on never-plowed land now sit ten feet higher than the surrounding cornfields, because of catastrophic erosion.

Topsoil is created very slowly. From a human perspective, it is essentially a non-renewable resource. It is highly challenging to grow grains or vegetables without causing topsoil loss, at varying rates. Logging and overgrazing also cause tremendous erosion. Practices that cause unnatural levels of erosion can be described as soil mining — they are done at the expense of the ecosystem and the generations yet-to-be-born.

Annuals & Perennials

The core of agriculture is based on growing a type of plants called *annuals* (i.e., corn, wheat, rice). Annuals die at the end of the growing season, leaving behind seeds from which next year's plants will grow. *Perennials* are plants that can live for more than a year (i.e., almonds, asparagus, blackberries, apples, redwoods).

On most farms, annuals are planted on plowed fields, which is not good for the soil. Some farmers are now using reduced-till methods (discussed later), which are also not problem-free. Perennials do not require annual plowing, which is an important advantage. Depending on the methods used, perennials can be grown sustainably or destructively.

In wild nature, annuals are emergency first responders, like firefighters or paramedics. When soil is disturbed or exposed by fires, floods, landslides, stampedes, or volcanoes, annuals are the first plants to grow, and they help hold the soil in place until more permanent vegetation can take over — similar to a scab forming on a skinned knee. Unlike most other plants, many annuals grow well in monoculture.

Perennials invest significant energy into roots and branches, to improve their chances for survival, year after year. Their roots grow deep, giving them access to nutrients that are unavailable to shallow-rooted annuals. Seeds are less important — an apple tree will survive even if it produces no seeds for several years. Perennials are the dominant vegetation on stable, uninjured landscapes.

Annuals have little need for big, deep root systems, and invest much of their energy into seed-making. Annuals often produce seeds that are packed with rich oils, proteins, and carbohydrates, which improve their chances for germination and survival. Because seeds contain so many nutrients, many animals seek them out, including humans.

Agriculture began in the Middle East, usually in the floodplains beside rivers, where seasonal floods often swept away the vegetation and exposed bare soil. Wild wheat, millet, and barley commonly grew on exposed soils that had been "plowed" by the floods.

In his book *Against the Grain*, Richard Manning speculates that the end of the Ice Age was a time of change and instability. In an era of melting glaciers and shifting climate, catastrophic floods were common. The wild ancestors of annuals like corn, wheat, and rice were plants that were highly adapted to thrive in the wake of catastrophe. The nutritious seeds produced by these plants attracted foragers and, in several locations, this eventually led to agriculture and civilization.

Thus, the core of contemporary agriculture was built on plants that do best in catastrophe. They thrive on injured land. On most farms, an artificial catastrophe is repeatedly provided by plowing, which eliminates the existing vegetation, buries the beneficial microorganisms, depletes the organic material, and leaves the precious soil exposed. The injured soil is never allowed to heal. You could say that the foundation and lifeblood of our lifestyle and culture is catastrophe.

The bottom line is that perennials build new soil when they are managed by nature or careful farmers. Annuals require tilling, and growing them inevitably depletes the soil. Soil depletion is the fatal defect of the Agricultural Revolution, and it has been a primary factor in the collapse of most civilizations. In their book *Empires of Food*, authors Evan Fraser and Andrew Rimas argue that the insatiable need for fertile soil has been the driving force behind imperialism and empire-building throughout history.

We do have the option of creating a form of agriculture centered on perennials instead of annuals. Perennial agriculture has the potential to be truly sustainable, if done well. Over the long run, with the return to a sustainable population, we have the option of reducing or eliminating our dependence on agriculture — the option that is most highly recommended by the spirits of our wild ancestors.

Watering the Crops

Agriculture requires water — no water, no food. There is no substitute for water. Crops can get their water from annual floods, rain, and/or irrigation.

Annual Flooding

Some farming is watered by annual flooding. The most notable example is Egypt, where the agricultural system lies in the Sahara Desert, along the Nile River. For thousands of years, annual floods delivered water, nutrients, and humus to the fields. Dams built during the 20th century have brought an end to the ancient mode. Additional discussion on the history of Egyptian agriculture can be found in *The Nile* section, beginning on page 166.

Rain-fed Agriculture

Some regions receive adequate rain during the growing season, avoiding the serious problems associated with irrigation. Rain-fed agriculture is not without problems.

- Rain is not 100% reliable. During droughts, crops are damaged or destroyed by inadequate rain. Droughts have caused many millions to starve over the centuries.

- In growing seasons when the rainfall is above average, plowing and harvesting can be hampered, fungal diseases are encouraged, and erosion increases, especially on sloped fields. The potato blight in Ireland is a good example of wet season problems.

- Intense rainstorms can flatten crops and cause erosion. These storms are common in the American Midwest.

Haiti is on the island of Hispaniola, in the Caribbean Sea. From June through November, heavy rains fall almost every night. This was not a problem when the land was protected by its original rainforest jungle. Today, only a tiny remnant of the forest survives, and it is being destroyed by charcoal-makers. The countryside has been plowed for agriculture, and this has resulted in catastrophic erosion. My geologist friend, Walter Youngquist, has been to Haiti, and he has never seen worse erosion — virtually no topsoil remains. Meanwhile, population is exploding, and their problems have been compounded by a severe 2010 earthquake. It is a land with a challenging future.

England is at the other end of the spectrum. Its soils have managed to survive the plow era better than most places — some fields have been in use for 1,000 years or more. Rainfall was adequate, and it generally arrived in the form of gentle showers. The soils were fertile, but not light and highly erodible. The landscape largely consisted of gentle rolling hills (fields on steep slopes are highly vulnerable to erosion). But, like Haiti, England's original forest cover is largely gone, the nation suffers from intense overpopulation, and the majority of its food is imported.

Irrigation

The use of irrigation in farming has been problematic over the centuries, which is a nice way of saying that almost all civilizations that utilized irrigation self-destructed over time. For thousands of years, irrigation has caused two serious problems: salinization and waterlogging. Today, these two problems have reduced or eliminated the productivity of one-third to one-half of the world's irrigated lands.

In modern times, we've added three new irrigation problems: over-pumping (taking too much water from lakes and rivers), groundwater mining (depleting underground aquifers), and dams. From 1950 to 2000, the area of irrigated land tripled. Dam-building spurred this increase, and then motorized pumps spurred it more.

Sandra Postel wrote *Pillar of Sand*, an excellent book on irrigation. Irrigation makes agriculture possible in places where there is little or no rain. Dryland soils are often quite fertile, because their nutrients haven't been leached out by centuries of precipitation. In warm regions, irrigation can enable up to three crops to be grown in a year. Just 17% of global cropland is irrigated, but irrigated farms produce 40% of the global harvest. In the US, 20% of grain is grown on irrigated land. In India, it's 60%, and in China it's 80%. Experts agree that irrigation is almost always more productive and dependable than prayers for rain.

Globally, 70% of all freshwater withdrawals are used for irrigated agriculture. Of this irrigation water, less than half is actually used by the crop. Most of the water is lost through canal leakage, spillage, infiltration, evaporation, and runoff. Today, 90% of irrigated land is watered by primitive flood and furrow systems, which are very similar to the irrigation systems that caused the extinction of many ancient civilizations.

Modern drip irrigation technology can reduce water usage by 70% or more, and increase yields by 20% or more, but it is expensive to install. The cost inhibits widespread adaptation — especially where water is subsidized and cheap. Israel is a leader in drip irrigation.

Dams

In many regions, the season of high precipitation (if any) does not correspond with the water needs of agriculture. In these places, dams have been built to collect water in reservoirs. This water is then used for irrigation. During extended droughts, the precipitation is sometimes inadequate to refill the reservoirs, and this limits the amount of water for irrigation.

Millions of people have been displaced by dam construction, because their homes were in locations that would soon be under water. Reservoirs often submerge high quality flood plain cropland.

The Klamath River flows from Oregon to California. Dams on the river have created warm, stagnant reservoirs that have turned into breeding grounds for toxic blue-green algae, which causes illness and liver failure.

Migratory fish like the salmon have lost access to most of their spawning grounds, because of dams like those on the Klamath River.

The salmon population has fallen substantially. Dams on the Nile destroyed the sardine fishery there.

Reservoirs in places like Egypt, Ghana, and Sudan provide excellent habitat for snails, which can carry schistosomiasis (a parasitic flatworm). In addition to snails, the reservoir behind the Aswan High Dam led to an explosion of mosquitoes, which spread Rift Valley fever, a disease that kills both humans and livestock.

All dams are temporary. Few will last for many centuries, and there is already a long list of ruptured dams. Many dams will eventually fill with silt and become useless waterfalls. Removing the accumulated silt from reservoirs is very expensive. The Sanmenxia Dam on the China's Yellow River was completed in 1960. By 1964, the reservoir contained huge amounts of silt, which reduced the water-storing capacity by 41%. Rapid siltation caused the Laoying Dam to be abandoned even before its construction was completed.

Waterlogging

Irrigation water that is not absorbed by plants will run off, evaporate, or drain into the aquifer. In places where the water table is close to the surface, irrigation causes waterlogging — the water level rises up into the root zone of the crops. Waterlogging reduces crop yields and promotes salinization — the buildup of salts in the soil.

State-of-the-art irrigation combines drip irrigation and drainage systems, which can eliminate or reduce waterlogging, salinization, and the waste of irrigation water. This technology is expensive to install, and is not used on most irrigated farms.

Like many great solutions, drainage systems also create new problems. The water drained from a field is likely to contain concentrations of nitrogen, salts, and agro-chemicals. Where do you send this contaminated water? It will cause problems wherever it goes, and the problems will grow each time the water is reused downstream. Thus, *state-of-the-art* should never be confused with *sustainable*.

Salinization

Dissolved salts are found in almost all water, including groundwater, raindrops, and surface water (i.e., rivers, lakes, oceans). In addition, saline soils contain mineral salts, and these soils are common in irrigated regions. In irrigated fields, the plants absorb the water, but not the salt. When water evaporates, the salt is left in the soil. Crops like cotton and sugar cane are water-guzzlers — the more water you apply to a field, the faster the salts will build up. Salt buildup is also encouraged by poor drainage (waterlogging).

Over time, the concentration of salt in the soil reaches a level that inhibits the ability of plants to absorb water. Heavy salt buildups make a field look white, like snow. No crops will grow in these fields. They are ruined. Many famous civilizations have been destroyed by salinization. Salinization often reaches a terminal state within 100 years of the start of irrigation.

Salinization is currently a serious and widespread problem, destroying large areas of irrigated cropland. It is likely to worsen because irrigation has been introduced in many locations in recent decades. The benefits of irrigation are short term. There are no long term benefits.

Water Mining

Following World War II, diesel and electric motors came into widespread use for irrigation. Affordable pumps were used to irrigate fields with water taken from rivers and wells. In many cases, the rate of pumping was not sustainable.

The most famous example of over-pumping is the Aral Sea in Russia. Water was pumped from the two rivers that fed into the sea, and used to irrigate a region the size of Ireland, so that cotton could be grown for a while on poor soils. Eventually, the sea shrank in size, its fishery died, the water became highly salty, the surrounding fields were white with salt, rates of cancer and lung disease increased, and the region's ecosystem and economy was devastated. The destruction of the

Aral Sea was planned and expected — the experts were not surprised by the results.

Rivers are being emptied by over-pumping for agriculture. At the same time, water usage for industrial purposes is also on the rise. For example, the Yellow River and the Colorado River often go dry before they reach the sea. Many big cities in China are experiencing serious water shortages.

Underground aquifers are also being over-pumped, in excess of their recharge rate. This is called groundwater mining. Over-pumped aquifers will eventually be emptied. Groundwater mining is happening in many regions, on a major scale. The Ogallala aquifer lays beneath a number of states in the central US, and it provides 20% of our irrigation water. It is being pumped dry, a third of it is gone, and everybody knows this. There are many similar stories in India, China, and California. In Saudi Arabia, aquifers were over-pumped, and the wheat harvest fell by half, between 1992 and 2004.

The Future of Water for Agriculture

Humankind's current water usage is likely close to it's historic peak — Peak Water. We are becoming more aware that the world's supply of fresh water has real limits. These limits are going to create growing problems over the coming decades.

- Heavy rains will continue to erode exposed topsoil. Climate change is likely to increase precipitation in some regions.

- Periodic droughts will continue, and may worsen because of the effects of climate change.

- In irrigated regions, salinization will permanently ruin croplands at an accelerating rate, because the area of irrigated land increased 700% in the 20th century (salt frequently becomes a problem after 100 years of irrigation).

- Silt continuously collects in many dam reservoirs, reducing the space available for irrigation water. Building new dams is an energy-intense process, and it will be hindered by the end of the era of cheap energy. Most sites suitable for dams already have dams.

- Over-pumping will empty many ancient aquifers and make irrigation impossible. Food production will be reduced or eliminated in these areas. Many millions of people are being fed with crops grown with water from aquifers that are being pumped out.

- Irrigation is competing with growing industry and growing urban populations for access to river water. Industry commonly makes more money with a gallon of water than a farmer does, and politicians are well aware of this. In some regions, river water is too polluted for use on fields, or even industrial use.

- The end of the era of cheap energy will make pumping more expensive. Rising costs will eventually force many farmers to reduce or discontinue irrigation.

- As aquifers are drained, irrigation wells are being drilled deeper. Near Beijing, some wells have to be 1,000 meters (3,280 feet) deep before hitting water. The deeper the well, the more it costs to pump. Eventually, the depth of the well will reach a point where the cost of pumping gets too high to continue.

- Climate change is melting high elevation glaciers and ice packs at a rising rate. Major rivers, like the Ganges, Indus, Brahmaputra, Mekong, Yellow, Yangtze, and Colorado, could become seasonal streams. This could disrupt irrigated farming systems that feed hundreds of millions of people.

- Fresh water can be extracted from salt water via desalinization plants, but this process is expensive and energy-intensive — the water produced is not cheap.

- Because of these issues, the cost of food will rise, and the supply will become less reliable.

- Conflicts over water are certain to escalate over the coming decades. There will be water wars.

- Current levels of agricultural production cannot be maintained. Our current system of agriculture is completely unsustainable.

- Water shortage is a bigger threat than Peak Oil. Water is needed for life, and there are no substitutes for it.

Feeding Our Green Relatives

Maintaining Soil Nutrients

Plant life requires soil, water, air, sunlight, and nutrients. Nutrients are extracted from the soil by the roots. Different plants have different nutrient requirements. For example, corn, rice, tobacco, and cotton consume above average quantities of nutrients. Of all the nutrients consumed by plants, the big three are NPK: nitrogen (N), phosphorus (P), and potassium (K). Animal life is also dependent on these essential nutrients. Eliminate N, P, or K from an ecosystem, and all life will disappear.

In smaller amounts, plants also need nutrients like calcium, magnesium, sulfur, iron, chlorine, copper, manganese, zinc, molybdenum, and boron. If a soil is deficient in any of these nutrients, plant growth and vitality are affected, or made impossible. Animals can also suffer from nutrient deficiencies. For example, osteoporosis is related to inadequate calcium.

Mother Nature, of course, does an excellent job of managing plant life. She encourages a diverse blend of vegetation that is appropriate for each location, recycles all nutrients, maintains soil health, and creates no waste — everything is used and reused, and no toxic sludge is sent to landfills.

Farming, on the other hand, is not an elegant balancing act. Every crop removes an assortment of nutrients from the soil, and nutrient recycling is not automatic — it's optional, and it's expensive. A trainload of wheat zooms away to New Jersey, and those nutrients are gone. The consumers of New Jersey do not display their deep appreciation by sending back a trainload of composted humanure. Instead, New Jersey will turn the wheat into bagels, feed it to humans, turn it into toxic sludge, and dump the sludge into a landfill. Somehow, the wheat field's lost nutrients must be replaced, or the farmer will not enjoy a happy future.

The depletion of nutrients in the soil is as old as agriculture. The rate of depletion varies by local conditions and by the farming methods used. In lean rainforest soil, a cleared field can be depleted in as little as a year or two. In the richer virgin soils of colonial America, it took three to ten years. In primitive farming, a field was used until crop yields diminished, then it was abandoned, and a new field was cleared. This worked as long as there was virgin soil for future use. Eventually, population pressure and repeated use of older fields made it harder for farmers to survive.

Over the centuries, farmers discovered some tricks for replenishing the soil nutrients. They added compost, manure, seaweed, crop residues, lime, ashes, fish scraps, slaughterhouse wastes, and many other things. They recycled nutrients from the local area. Although this was labor-intensive, it reduced the need to abandon depleted fields. Recycling is never 100% efficient — a portion of the nutrients are lost. Nitrogen is most easily lost.

The guano industry started about 1820, when white men discovered that guano (bird, seal, and bat poop) was a high-potency fertilizer (Peruvian Indians had known this for centuries). Guano contained significant amounts of nitrogen, phosphorus, potassium, and other nutrients — it contained 30 times more nitrogen than manure.

There were rocky cliffs and islands in South America where the guano had been collecting for thousands of years, and in some places it was piled into mountains 200 feet deep. Capitalists fell over each other in the race to get rich in the guano gold rush. The precious guano was mined, loaded onto boats, shipped to distant lands, and sold to farmers. The industry thrived from 1840 to 1870, until the treasure chest of poop ran out.

The guano era brought a big change to agriculture. Farmers were no longer forced to recycle local nutrients. They now had the option of buying nutrients. The use of manure became optional, so the need for devoting land to manure-producing livestock diminished. Farmers who were too poor to build a barn and run a livestock operation could now go to town and buy fertilizer.

As the guano supply waned, the fertilizer industry shifted to acquiring nutrients from mineral sources — sodium nitrate (N) from Chile, potash (K) from Germany, and phosphate (P) from England, France, and Spain.

Today, nutrients are replaced by applying chemical fertilizers. This system is extremely energy-intense, and absolutely unsustainable. There will be big problems in the fertilizer department as we move beyond the era of cheap energy.

Nitrogen (N)

Every plant and animal must absorb nitrogen in order to survive — no nitrogen, no life. The Earth's atmosphere is 78% nitrogen, but it is not in a form that plants or animals can absorb. Atmospheric nitrogen consists of pairs of nitrogen atoms (N_2).

Nature converts nitrogen into a usable form via bacteria in the soil. The bacteria break apart the pairs of atoms, and join the nitrogen atoms to hydrogen atoms, creating ammonia molecules (NH_3) that living organisms can absorb. This process is known as *nitrogen-fixing*. Nitrogen-fixing bacteria grow on the roots of *leguminous* plants, which include peas, beans, clover, and vetch. Thus, legume plants add nitrogen to the soil, and other types of plants extract the nitrogen (corn, wheat, tomatoes, etc.). Plants absorb the nitrogen from the soil via their roots, and animals absorb nitrogen by eating plants or other animals. Legumes provide more nitrogen to the soil than manure does.

In order to maintain crop production, farmers must replace lost nitrogen. In the old days, there were four ways for doing this:

- *Composting* recycled local resources like animal manure, human wastes, and crop residues.

- *Crop rotation* was used. For example, a field might be used to grow corn one year (using up nitrogen), and the next year planted with nitrogen-fixing leguminous crops, like beans, peas, lentils, or soybeans (replenishing the nitrogen).

- Following a crop harvest, the field could be planted with a leguminous *cover crop*, like clover or vetch. This protected the soil from erosion during the off season, added nitrogen to the soil, and was then plowed under in the springtime, as a *green manure*.

- *Fallowing* was letting a field rest for a year or more. This practice became obsolete with the development of crop rotation and cover crops, which made more productive use of the cropland.

For a few decades, farmers had the guano option, until that resource was used up. An alternative nitrogen source was sodium nitrate, a mineral mined in Chile. Sodium nitrate was used from 1830 into the 1920s.

Agricultural productivity is restricted when there is an inadequate supply of N, P, or K available. Most commonly, nitrogen is the essential nutrient in short supply, and this shortage creates a bottleneck on yields. Nature only produces so much usable nitrogen, via nitrogen-fixing bacteria. The limited supply of fixed nitrogen puts a ceiling on the human population, in theory. Norman Borlaug, the father of the Green Revolution, theorized that relying solely on nitrogen-fixing bacteria would allow the existence of no more than four billion people.

The population ceiling was broken by two German lads, Fritz Haber and Carl Bosch. Haber figured out how to take nitrogen from the air, and combine it with hydrogen, using high temperature and pressure (he later invented Zyklon B, the poison used in Nazi gas chambers). The end product was synthetic ammonia (NH_3), a substance that plant life finds to be extremely nutritious — it was very potent. Carl Bosch figured out how to perform this magic on an industrial scale. Haber and Bosch opened the first ammonia plant in Germany in 1911.

Both lads were awarded the Nobel Prize, and many consider the Haber-Bosch process to be the most significant achievement of the 20th century. Richard Manning wrote that following the invention of the Haber-Bosch process, the fertilizer industry was able to produce as much usable nitrogen as Mother Nature did (via nitrogen-fixing bacteria) — doubling the world's supply of nitrogen for plant growth. There were 1.7 billion people in 1910, and now there are seven-point-something billion. Nitrogen expert Vaclav Smil speculated that 40% of the people alive today would not exist without Haber-Bosch.

Manmade ammonia replaced the sodium nitrate from Chile as a feedstock for explosives. Many new ammonia plants were built to make explosives for World War II. At the end of the war, large quantities of ammonia became available for producing ammonia fertilizer — the spark that detonated the population explosion, according to Smil. Following the war, fertilizer use skyrocketed.

Corn is a crop that is especially responsive to higher doses of nitrogen. Ammonia fertilizer can put a cornfield into fast forward. The following table shows average grain corn yields for the United States. Note the leap after 1945 (the end of World War II). The data is from the USDA.

Year	Bushels per acre
1885	28.6
1905	30.9
1925	27.4
1945	33.1
1965	74.1
1985	118.0
2005	147.9

By 1965, ammonia fertilizer was in widespread use. It had eliminated the need for crop rotation, so corn could be grown on the same field, year after year (instead of two out of five years). Some of this growth was also due to the development of hybrid corn varieties, which were highly responsive to increased nitrogen.

These statistics also suggest what we can expect for yields in the less-productive future, as we move beyond the era of cheap energy and cheap fertilizer. Currently, most ammonia fertilizer is made using natural gas. Making nitrogen fertilizer consumes 2% of the energy used in the world. Ninety percent of the cost of making ammonia fertilizer is the cost of natural gas. Because of the cost of gas, the US imports more than half of its ammonia. China and India lead the world in ammonia production.

Natural gas is a finite non-renewable resource that is subject to sudden shifts in price. For example, here are USDA figures for the cost of ammonium nitrate in recent years:

Year	Dollars per Ton
2002	$195
2003	$243
2004	$263
2005	$292
2006	$366
2007	$382
2008	$509

Natural gas is not needed to make ammonia. In China, 60% of the nitrogen fertilizer is made using coal. Ammonia made using natural gas or coal is one of the biggest single sources of greenhouse gas emissions. Ammonia can also be made using hydropower, but the cost is many times higher.

Some predict that our population will hit 10 billion by 2050. More than a few experts believe that feeding that many people is simply not possible. We would need to double ammonia production in order to feed them, according to Stan Cox, author of *Sick Planet*. Cox has developed an ambitious plan for deliberately eliminating the use of nitrogen fertilizer in the US. He recommended that we:

- Follow nutritional guidelines for meat consumption, and eliminate excessive meat eating.

- Raise livestock on pasture and rangeland, and stop feeding them grain.

- Rotate grain crops with non-harvested crops of nitrogen-fixing legumes.

- Reduce the cropland used for growing annual crops, and replace it with deep-rooted perennial plants.

- Reduce food waste. Cox suspects that 25% of the food produced is discarded. Some estimates of US food waste are over 40%.

- Cut grain exports.

Phosphorus (P)

Providing phosphorus is less complicated than nitrogen — it doesn't require processing by soil bacteria (or industrial chemical plants) to turn it into a usable nutrient. But providing phosphorus also has some challenges. It isn't readily distributed through the soil by the rain, it disperses very slowly. It works better when the fertilizer is tilled into the soil, rather than spread on the surface. The tricky part is that if the soil is not slightly acidic, the phosphorus will be locked up in a form that cannot be used by the plants. Many organic farmers have trouble with phosphorus deficiencies.

In earlier days, phosphorus was recycled. Animals excrete most of the phosphorus that they consume, so returning urine and manure to the fields, along with compost, returned phosphorus to the soil. There is a great deal of phosphorus stored in sewage treatment plant lagoons, but it is very expensive to extract, because it is mixed with many chemicals. So, this source of phosphorus is turned into toxic sludge and sent to a landfill. It's much easier to recycle phosphorus when it is fresh from the outhouse.

From about 1840 to 1870, guano was a popular fertilizer that provided a source of nitrogen, phosphorus, and potassium. Bone meal was also used to provide phosphorus to fields. People would go to Napoleonic battlefields, gather the bones of the dead, and turn them into fertilizer. The Nazis made fertilizer from the bones of those who were killed at concentration camps. The bones of exterminated American bison also fueled a brief but profitable industry.

In modern industrial agriculture, most farmers do not fertilize their fields with manure, urine, or bones, for economic reasons. The source of phosphorus used today is the mineral compound phosphate. In the second half of the 19th century, phosphate mining began in a number of locations. Powdered phosphate rock could be spread on fields to provide the needed phosphorus. Today, phosphate mining is done on a large scale.

In recent years, the phosphate content of the mined rock is declining. The lower-quality rock contains higher levels of cadmium, arsenic, and other heavy metals, which are concentrated during processing. Not all of the toxics are removed by processing — a portion end up in the fertilizer sack. Production byproducts include radioactive gypsum and airborne fluorides. In the coming years, the quality of the rock is expected to decline further, and the cost will rise.

Some experts are predicting the peak of phosphorus production to occur by about 2040, others say 2033, others say sooner. The cost of phosphate rock is sharply rising. Reserves may last for another 50 to 130 years. There are phosphate mines in 30 countries, but 37% of reserves are in China, and 32% are in Morocco. China recently put restrictions on phosphate exports — a 135% tariff, which alarmed a lot of people. Phosphate is a strategic mineral, and its distribution could be controlled for political purposes. American phosphate production is past peak, and in decline.

Potassium (K)

Potassium is a nutrient that is easy to recycle. Animals excrete most of the potassium that they consume, so returning urine and feces to the fields is highly beneficial to soil fertility.

Potash is the chemical compound used in commercial fertilizers to provide potassium for plant growth. In an earlier era, sources of potash included tobacco stems, wood ashes, cotton hulls, and cotton boll ashes. If you burn 1,000 tons of wood, the ashes will contain one ton of potash. In 1857, Germans began mining and processing mineral potash. World reserves of mineral potash are immense. Potash is also used to make soap and gunpowder, but the fertilizer industry consumes 93% of potash production.

The biggest deposits are located in Saskatchewan, Canada. We'll run out of many crucial resources before we use up the potash rock.

But, as the era of cheap energy concludes, it's going to be more and more expensive to mine, process, and transport potash.

Lime (Calcium Carbonate)

Lime is used to control the acidity of soil. Acidity is measured on the pH scale, which has 14 points. A pH of 0 (zero) is highly acidic. At the other end of the spectrum, a pH of 14 is highly alkaline, like lye. A pH of 7 is neutral.

As a general rule, drier climates tend to have alkaline soil, and wetter climates often have acidic soils. In wetter climates, the rains can leach out the chemicals that counteract acidity, resulting in higher acidity. Lime must be periodically reapplied to keep the acid level under control. When the soil is too acid or alkaline, some nutrients do not dissolve effectively, which inhibits plant health, and some toxins are converted into a form that is available for absorption by plants.

In the old days, farmers used wood ash, crushed seashells, and other substances to provide lime. Today, crushed stone is used to provide lime: limestone or dolomite. Neither mineral is in short supply, but the supply of both is finite.

Significant quantities of mineral lime are produced every year. This lime must be mined, crushed, transported, and applied to the fields. Each of these processes requires energy and equipment. As we move beyond the era of cheap energy, the cost of lime will rise, and the supply may become unreliable. This will have a negative effect on crop yields.

Adventures in Sewage

In a fascinating essay titled *Civilization & Sludge*, Abby A. Rockefeller described the evolution of sanitation systems. The following is mostly a summary of her essay.

Rockefeller learned that the simplest and most sustainable sewage treatment system was developed by nomadic foragers. They utilized the same time-proven system used by non-human animals — depositing their feces and urine on the ground, in a widely dispersed manner. This recycled vital nutrients, cost nothing, required no staff or infrastructure, did not pollute the water, kill the fish, encourage the spread of contagious water-borne diseases, or produce a single spoonful of toxic sludge. This brilliant system works very well in societies having low population density.

With the advent of agriculture, the supply of food increased, the population increased, the output of sewage increased, and the old system failed completely. This inspired the clever invention of smelly outhouses and cesspools. This new technology recycled nutrients less effectively than the nomadic forager system.

The flush toilet grew in popularity during the 19th century, as municipal water systems came into fashion. Municipal water systems increased the production of wastewater, which overwhelmed the old cesspools. The cheap and dirty solution was open sewers — ditches beside the streets where sewage from the cesspools was drained. It's no coincidence that cholera became a very popular disease at this time.

This inspired the development of closed-pipe sewage systems, which moved the wastes out of town — into lakes, streams, and oceans, where nature would (in theory) purify it all. On the plus side, cholera rates dropped. On the downside, typhoid became popular among downstream residents who got their water from sewage-laden streams. Once upon a time, the Thames River of England was filled with salmon, and supported a thriving fishery. Then came the new and improved sewage systems, which killed the fish, and turned the Thames into one of the most polluted rivers on Earth.

This inspired cities to filter the drinking water pumped from tainted waterways. Typhoid rates dropped. But filtering did not remove the sewage from the rivers, and rapid growth in the industrial sector was adding large quantities of other pollutants, including toxics.

This inspired cities to treat waste before dumping it into waterways. Treatment systems have been evolving over the years — each new design is more complex, expensive, and energy-intensive than its predecessor. The wastes and nutrients that used to go into the river are now concentrated into toxic sludge.

Because the waste discharges from industry varied from location to location, and day to day, the toxicity of the sludge varies from moderate to extremely poisonous. The sludge was dumped into the ocean, where the poisons created dead zones on the ocean floor. Ocean dumping was outlawed in 1988. At this point, sewage industry propagandists began presenting toxic sludge as a wonderful fertilizer — *beneficial biosolids!* This was given to farmers free of charge.

Toxic sludge is low in nitrogen, so it has to be applied in large quantities to serve as fertilizer. Heavy metals and other toxins in the sludge move into the soil. These toxins are absorbed by plants, and the animals that eat them. In the soil, thousands of industrial chemicals can interact, creating a countless opportunities for unintended and undesirable consequences.

Following the application of toxic sludge at a Georgia dairy farm, the milk was contaminated with high levels of toxic thallium. Another Georgia farmer watched his herd of 300 cattle die — his free beneficial biosolids happened to contain high levels of arsenic, heavy metals, and PCBs. Sludge is a hazardous waste. What do we do with it? Answer: stop making sludge. Human wastes need to be returned to the soil, and production of toxic industrial wastes needs to end.

What is the moral of this story? Thou shalt keep society small and simple. Ants and bees live well in large complex civilizations. But humans are not insects. This is an important fact to remember.

Blue Babies & Dead Zones

Fertilizer is a fascinating thing. When it is applied to farmland, crop yields grow, population grows, the economy grows, resource consumption grows, pollution grows, conflict grows, and life becomes miserable — progress! In his book, *The New Green History of the World*, Clive Ponting documented a sharp increase in the use of synthetic fertilizer in the world (a growth curve quite similar to world population). You can see that fertilizer use skyrocketed after World War II.

Year	Tons Used
1900	360,000
1950	10,000,000
1990	150,000,000
2000	137,000,000

There are limits to the benefits of fertilizer. For example, from 1910 to 2000, fertilizer use in Western Europe increased ten times, while crop production merely doubled. In China, nitrogen and phosphorus usage increased ten times between 1949 and 1983, while yields only tripled.

Estimates suggest that 40% to 75% of the fertilizer used does not make contact with the intended crop plants. Some of the unused fertilizer is stored in the soil, but most of it filters down into the groundwater, or gets washed into waterways, where it does an excellent job of fertilizing things that we didn't intend to fertilize, creating serious unintended problems, like dead zones. Also, the chemicals in fertilizer can react with other chemicals in the environment to produce noxious chemicals. Thus, the dark side of miracle fertilizers is that they are also powerful pollutants. Some perceive nitrogen fertilizer to be a water pollutant that is also used to nourish crops.

Polluted groundwater is often used as a source of drinking water. Some experts worry that nitrates in our water could be associated with kidney and urinary tract problems, and that they may be carcinogenic. Nitrate levels in underground aquifers can accumulate with time, and persist for centuries. Nitrate levels in the water supply for the city of Des Moines, Iowa (corn country) are at record levels. The city spends a lot of money filtering the nitrates out of the water.

Nitrates can break down into nitrites. Nitrites bond with hemoglobin, and this reduces the ability of blood to carry oxygen, which can lead to serious life-threatening problems like blue baby syndrome. Nitrates can also break down into nitrosamines, and these have been linked to cancer in lab animals. Another byproduct is nitrous oxide, a powerful greenhouse gas.

Fertilizer runoff encourages the growth of aquatic plants, which leads to more snails, which leads to more schistosomiasis. Runoff also inspires population explosions in algae. The algae suck up the oxygen in the water, and this leads to dead zones. Fish swimming into a dead zone lose consciousness and die of suffocation. In 2008, there were over 400 dead zones in the world, and they are increasing at a healthy rate. One of the world's largest dead zones, located in the Gulf of Mexico, is the size of New Jersey. It was killed by fertilizer runoff from America's corn-growing heartland.

On the bright side, some dead zones can rise from the dead. The Black Sea dead zone disappeared following the collapse of the Soviet Union, when fertilizer use sharply declined. Other dead zones are more enduring, like those caused by dumping toxic sludge into the ocean.

We worry a lot about super-pollutants — stuff like dioxin, PCBs, mercury, radiation. We need to worry about fertilizer just as much — and maybe more — it's still legal and in heavy widespread use. Its intended "benefits" enable overpopulation, and promote soil mining, water mining, and other problems. Its unintended consequences are also huge problems.

Nutrient Recycling

There will come a day when we will no longer be able to ride into town and buy synthetic fertilizer, because its production and distribution requires the consumption of ever-diminishing finite resources. The days of amazing fertilizers and amazing yields are going to fade away into bad memories of an era of exploding population, blue babies, dead zones, endless wars, and assorted catastrophes.

The heyday of powerful synthetic fertilizers may not last 100 years (1950-2050?). There will come a day when localized nutrient recycling once again becomes the primary source of fertilizer for all fields. This was the norm, prior to the 19th century, when there were fewer than a billion people on Earth, most of whom were undernourished.

In 1909, Franklin Hiram King took a trip to Asia. He was a professor and the Chief of Soil Management at the US Department of Agriculture. The purpose of his visit was to learn how the farmers of Asia produced so much food per acre, and the techniques that allowed some regions to be farmed continuously for 4,000 years. He was deeply impressed by what he saw, and he published a glowing account of his findings, titled *Farmers of Forty Centuries*. In a nutshell, maximum production was achieved by a combination of intense manual labor and the elimination of waste, via 100% nutrient recycling.

King was spellbound by productivity and efficiency, and his mind was a gushing fountain of numbers. He paid little attention to the miserable lives of poor peasants, and the ecologically-ravaged lands where they lived. This book describes a densely populated society of impoverished subsistence farmers, on postage stamp farms, who had large families, where everyone worked seven days a week, at hard physical labor. Their system was maxed-out, operating at 100% of capacity, just one step ahead of disaster — usually. Surpluses were minimal, and floods and droughts took heavy tolls. There were no social safety nets.

The secret to soil fertility was that all organic material was recycled. King observed the daily caravans of a hundred peasants pulling a hundred handcarts, each carrying 60 gallons of fresh sewage, from the city to the farmers' fields — a profitable, beneficial, and somewhat smelly industry. In the Yellow River region, nutrient-rich yellow silt was regularly dredged out of the irrigation canals and dumped on the fields.

Manure was sacred treasure. Horse poop did not sit on the street for long before a lucky lad found it and took it home. Barnyards were spotlessly tidy. They didn't buy mineral fertilizer shipped in from faraway places, because they couldn't afford it, they didn't need it, and they thought it was utterly stupid. Why buy what you can make?

King realized that the American and European addiction to imported mineral fertilizers was a huge vulnerability for the long term, because the supply of minerals was finite, and using them was therefore unsustainable. The Asians thought that the American system was ridiculous — gathering excellent fertilizer in centralized sewage systems, and then dumping it into a lake, stream, or ocean (often a source of drinking water) — polluting the aquatic ecosystem — and then paying good money to buy fertilizer from distant lands. Ridiculous indeed!

Sewage recycling in Asia diminished in the 20th century, as population increased, cities grew in size, and the journey from town to field got too long. A number of rapidly-growing cities in China are now dumping raw sewage into the rivers.

There are few pathogens in urine, if it is kept separate from feces. It can be mixed with some water, and taken directly to the garden. Urine is a fabulous fertilizer, containing lots of nitrogen. Urine contains 80% of the nutrients that we excrete. Unfortunately, urine is usually mixed with feces. Fecal bacteria break down the urine, causing much of the nitrogen to be lost. When urine and feces are not mixed, there is far less odor.

Manure can be dried, but drying results in a significant loss of nitrogen. Dogs and hogs find human feces to be totally delicious. Some believe that this irresistible flavor experience was a primary factor in why they became domesticated. In China, the outhouse and pig pen were often back to back.

Efforts were sometimes made to sterilize the sewage by composting it, but these were not entirely successful. Under ideal conditions, composting can kill most of the pathogens. Joseph Jenkins wrote the bible of sewage composting: *The Humanure Handbook*. For safety's sake, he recommends a patient approach — let the feces compost for two years before using it. Humanure can also be put into fish ponds and rice paddies to provide nutrients.

Because they used fresh, un-composted human feces for fertilizer, most Chinese had worms, and many died from fecal-borne infections. The disease factor is a major reason why people drink tea in Asia (and avoid eating raw vegetables or fish) — boiling eliminated much of the danger. Thus, people who drink tea are more likely to survive to adulthood and reproduce. Coffee drinking provides the same benefits.

Alcohol is also a traditional disease killer. Alcohol kills germs because it is a poison. People who drink beer or wine are more likely to survive than those who quench their thirst with sewage-flavored water. When consumed in moderation, alcohol is less lethal than cholera or typhoid. Over time, the people in alcohol-drinking cultures evolved genetic resistance to the poison. Native Americans are not alcohol-tolerant, which is why their communities have been hammered with alcoholism.

Peak Fertilizer

The end of the era of cheap energy will bring many challenges. One of these challenges is that the production of synthetic fertilizer will reach its peak, and then decline. This will have a number of serious effects:

- The cost of the natural gas used to make ammonia fertilizer will rise, and the supply will eventually diminish, following the peak of gas production in the coming decades. Ammonia can also be made using coal.

- The reserves of phosphate minerals are also not abundant. Phosphate mining will peak and decline during the 21st century.

- The cost of mining, processing, and distributing phosphate, potash, and lime will rise.

- As synthetic fertilizers are phased out, and replaced with less-potent recycled local nutrients, crop yields are likely to fall substantially. Fields will no longer be overdosed with N, P, and K, and locally-produced fertilizer may be short on one or more crucial nutrients.

- The transition to nutrient recycling will be bumpy, because industrial farming has destroyed many of the topsoil microorganisms that process nutrients into usable forms. Chemical fertilizers deliver nutrients in a usable form, but recycled nutrients require processing by the soil organisms that thrive in healthy topsoil.

- A single farmer will no longer be able to run a 3,000 acre farm alone. There will be a growing migration of people from urban areas to farm country, to perform the labor that is currently done by machines and cheap petroleum.

- As food production declines, so will the population.

Weeds, Pests, and Pathogens

Tilling the Soil

Weeds are plants that grow in places where they are not wanted. Weeds reduce crop yields by competing with the crop plants for sunlight, water, and nutrients. Weedy fields are more likely to be damaged by pests and diseases. If weeds are not controlled, they would grow to full size, and the crop yield would be minimal. Gardeners can remove weeds by hand, or obstruct them with mulch, but farmers fight weeds with tilling or herbicides, both of which have drawbacks.

The benefits of tilling include:

- Burying young weeds and weed seeds (while at the same time bringing older weed seeds back near the surface, where they are more likely to germinate).

- Disrupting the nests of rodents, insects, and other pests.
- Exposing dark soil, which more readily absorbs solar heat, a benefit in regions with cold winters.
- Working fertilizer into the soil, where it will be more available to plants, and less likely to be washed away.
- Pulverizing the soil, making it easier for the root systems of crop plants to expand.

The drawbacks of tilling include:

- Soil is exposed to wind and rain, with no protective cover of vegetation. This encourages the soil to blow away and wash away.
- Pulverized soil promotes evaporation in some climates. In irrigated fields, more water has to be used.
- Tilling after the autumn harvest can result in a muddy springtime, because the loosened soil can absorb more moisture. Planting must be delayed until the soil dries out.
- Tilling releases the carbon dioxide stored in the soil, allowing it to disperse into the atmosphere. The loss of carbon dioxide eventually destroys the soil's organic matter, which promotes erosion, and depletes the soil's fertility.
- Tilling buries the community of beneficial organisms that live in the top two inches of the soil, which destroys or disrupts their beneficial activities. This loss makes crops highly dependent on fertilizer for their survival.
- Operating heavy machinery in fields compacts the soil, which can affect crop yields.
- Heavy machinery consumes a lot of fossil fuel.

- Plowing and cultivating reduces weed problems, but does not eliminate them. As weeds emerge, conventional farmers often control them with one or more applications of herbicide.

- Herbicides are not 100% effective. Weeds commonly develop resistance to them. Herbicides are not harmless to ecosystems, and they are expensive.

It is difficult to imagine that tilling or herbicides can be components of a sustainable future. But weeding by hand is sure to enjoy a healthy revival.

The Dust Bowl

In the 1920s US farmers bought tractors and began plowing up the short grass prairies of the Great Plains — a dangerous thing to do in a dry region with highly-erodible loess soil. The first storm of the Dust Bowl began on November 11, 1933, and within 24 hours some farms had lost all of their topsoil. In 1934, Congress members in Washington, DC were stunned to look up and see a sky darkened with western soil — which eventually traveled as far as Norway. The Dust Bowl destroyed 35 million acres of land, and severely damaged another 125 million acres.

In *Cadillac Desert*, Marc Reisner wrote that dust storms were similar to avalanches. The racing dust in the wind kept scouring up more and more dust from the ground, and the whirling mass of airborne dirt rapidly grew until the dust storm was a huge black cloud, several thousand feet high, containing millions of tons of soil particles. When the dust storm passed, the chickens would have no feathers, and houses had the paint sandblasted off of them.

All farmers understand the relationship between tilling and erosion, and most understand the basics of erosion control. Unfortunately, pursuing state of the art soil conservation presents a serious obstacle to making enough money to pay the mortgage, bills, and taxes — because farmers do not control the prices for their crops, and the government remains unwilling to step in and vigorously defend the wellbeing of the generations yet to be born. The dominant goal of the food industry is maximizing profits by any means possible.

In the US, soil conservation programs have come and gone, erosion rates are lower than in the past, but the problem remains huge, irreversible, and very serious. Parts of northern China, Mongolia, and Africa are currently experiencing catastrophic erosion. Soil from Chinese dust storms is being deposited in western North America. African dust storms are sending soil to Florida.

More than a few learned elders have come to the conclusion that the plow has caused more harm than the sword. Replacing lost soil takes geological time. From a human perspective, soil is a non-renewable resource. The damage caused by erosion is permanent.

Conventional Reduced-till Farming

"No-till" farming is a new method that requires less tilling, but the name implies that no tilling is done, which is false. A less common, but more accurate name is *reduced-till*. In an effort to minimize erosion while maintaining yields (and to become eligible for additional government subsidies), some farmers are switching to reduced-till methods. In 2004, reduced-till was being used on less than 7% of global cropland.

On a farm that uses reduced-till methods, the field is not plowed prior to planting. A seed drill is used to cut slices in the soil, while planting seeds, and injecting fertilizer. Herbicide is used to kill the weeds. The benefits of reduced-till include:

- Erosion is significantly reduced, because the mat of dead vegetation from the previous season is left in place.

- Evaporation is reduced, so less irrigation water is needed.

- Because the need for plowing, disking, and cultivation is reduced or eliminated, fuel consumption is cut by 50% to 80%.

- Labor invested in producing a crop is reduced 30% to 50%.

The drawbacks of reduced-till include:

- Conventional till farming uses herbicides, but reduced-till uses more. Reduced-till is not used much in Europe, because of government controls on the use of agricultural chemicals. These chemicals are not ecologically harmless, and the long-term effects of their usage are not thoroughly understood.

- Reduced-till can only work as long as herbicides are effective. Over time, weeds routinely develop resistance to herbicides.

- Tilling exposes dark soil, which is more readily heated by the sun in the spring. Pulverized soil also dries more quickly. In cooler regions, like the Upper Midwest, seeds planted in reduced-till fields can rot, because the soil is too cool and wet.

- The conversion from till-based farming to reduced-till can be challenging and problematic. For example, the higher moisture content in the soil may promote fungal diseases.

- On some types of soils, yields can be reduced up to 10%.

- Reduced-till does not work well in regions having heavy clay soils.

- Up to 20% more nitrogen fertilizer is needed during the first four to six years following a conversion to reduced-till.

- Reduced-till does not work well when corn is grown year after year on the same field.

- Agribusiness corporations love reduced-till, because it continues to utilize farm machinery, petroleum fuels, chemicals, and proprietary seeds.

- Reduced-till still injures the soil.

Organic Reduced-till Farming

With the goal of eliminating herbicides from reduced-till farming, the Rodale Institute has been experimenting with organic reduced-till techniques. Following the harvest of the cash crop, the field is tilled and a cover crop is planted. When it is time to plant the cash crop, the off-season cover crop is rolled and crimped, which kills the plants. The cash crop is planted at that time.

This system is more challenging than the conventional mode, and it is not yet ready for widespread adaption. Success depends on two variables: the cover crop must be forming seeds at the ideal time for planting the cash crop, and the cover crop must be rolled and crimped before their seeds are fully ripe. So, the type of cover crop selected is a crucial factor, as is the timing of the planting. Stan Cox added that success requires perfect weather and lots of good luck.

Springtime tilling is still needed every two to three years, to control weeds. Tilling is also done every year, prior to planting the cover crop. This organic system still requires machines and petroleum, so it can never be considered sustainable agriculture.

Plow-free Alternatives

It is important to remember that agriculture did not exist for most of human history. It is not necessary to plant grains and vegetables every year. There are many ways to acquire healthy foods that do not require plows or tractors. Alternatives will be discussed in later sections. The bottom line is that there is no way to feed seven-point-something billion people in a manner that is not ecologically catastrophic. There are too many people today, but this will change.

Threats to Food Security

In the old days, almost all of the food that we ate came from the local area. Villages were fairly self-sufficient, and there was little travel from region to region. A low mobility way of life inhibited the spread of plant diseases, exotic weeds, and pests. Today, the foods on our dinner plate may have originated in several different continents. The high mobility of our global economy increases the risks to food security, as weed seeds, diseases, and pests are shipped from region to region via ship, train, plane, and truck.

For example, in 2004, farmers near Baton Rouge, Louisiana were shocked to observe that their soybean fields had suddenly become defoliated. The Asian soybean rust fungus spread to eight other states within weeks. It prefers cool, moist weather, and it has been known to destroy 80% of a crop.

In 1845, the potato blight moved from America to Ireland, causing a catastrophe that resulted in a million deaths. In 2003, potato blight spores arrived in Papua New Guinea and wiped out the crop. The blight fungus remains the biggest threat to potato crops, and it never stops mutating. Plant breeders and fungicide chemists work continuously to create temporary solutions to the latest mutations. Blight is a never-ending threat, wherever the spud is grown, and it can completely wipe out the crop across wide regions in a matter of days. A single field can turn black overnight.

A century ago, a fungal disease called stem rust periodically destroyed 20% of America's wheat harvest during wetter years. It has been damaging wheat crops since the days of Rome. In the 1950s, Norman Borlaug, the Green Revolution's famous plant breeder, identified a single gene that provided resistance to stem rust, and he created a new variety of wheat that carried that gene. The gene also improved yields, so it was rapidly adapted by farmers.

In 1998, a plant breeder in Uganda observed stem rust, which hadn't been seen in decades. Many believed that the fungus had gone extinct. The fungus had mutated into a form named UG99, and almost all of the wheat grown in the world today has no resistance to it. Under ideal conditions, it can destroy 100% of the wheat in a region. Its spores can hitch rides on the clothing of international travelers. The spores can travel with the winds.

As I am writing, it has spread to eight nations, and it is approaching the Punjab, a major wheat-growing region. Many experts believe that it will inevitably spread to the Americas. During the Cold War, other strains of wheat rust spores were stockpiled as biological weapons by the Soviet Union and the United States.

It's a similar story in the world of bugs. I'd like to introduce you to *Helicoverpa armigera*, a bug that has gained the reputation of being the world's worst insect pest (also known as the cotton bollworm, the American bollworm, the corn earworm, the tobacco budworm, etc.). This bug has only become a problem because the predators that usually eat it are highly vulnerable to pesticides. It is not a fussy eater, and it delights in feasting on many crops, causing billions of dollars of damage every year. It is resistant to almost every type of chemical insecticide.

The history of the spread of exotic diseases, weeds, and pests makes fascinating reading for those who enjoy scary stories. None of our primary food sources are safe and secure. High mobility will always be a deadly threat to civilization, because it is constantly moving problem-causing genes and organisms from continent to continent.

In the old days, farmers sowed their fields with good old-fashioned open-pollinated seeds from last year's harvest. This resulted in healthy genetic diversity in the field, because every plant had a unique mix of genes (like every human has a unique fingerprint). Diversity provided some protection from diseases and pests. Fields planted entirely with identical genetic clones, like the hybrid varieties commonly used in industrial agriculture, are far more susceptible to disaster — what can kill one plant is more capable of killing them all.

An estimated 140,000 varieties of rice have been developed in Asia, but the number of varieties used in mass production today is small. For example, in Pakistan, 80% of the cropland used for rice production is planted with a mere five varieties. Some plant disease epidemics, like the rice blast fungus, are capable of destroying 100% of a crop. Disaster is always just one mutation away, and genes are mutating in every type of organism every day.

In the old days, subsistence farming was the norm. People grew most of the food they ate, and each farm produced a variety of foods. Today's farms grow monocultures of commodity crops, and farm families buy their food at Safeway. The risks of planting genetic clones are multiplied when the clones are planted in vast monocultures. This system presents huge opportunities for trouble. Monoculture farmers can suddenly lose everything. Subsistence farming was less risky. A small village surrounded by orchards, nut trees, vegetable gardens, chicken coops, grain fields, and pig pens is far safer than a thousand acre field of genetically identical corn plants.

Virtually every farmer — both organic and conventional — experiences varying degrees of crop damage due to weeds, diseases, and pests. Industrial agriculture responds to these challenges by applying herbicides, fungicides, insecticides, and rodenticides. Mother Nature responds to these toxic chemicals by producing chemical-resistant organisms via the magic of mutation. A wonder-chemical may kill 99% of the target, but the chemical-resistant 1% will survive, reproduce, and present a new, and possibly more threatening, challenge to industrial agriculture. Mother Nature is winning. Nature will always win in the long run. Smart humans strive to live with nature, rather than control it.

When weeds or insects become chemical-resistant, the farmer can switch to a different chemical, and use it until the target develops resistance once again. Because nature is always mutating, no chemical provides permanent control. So, the development of new herbicides and insecticides is a non-stop process. In 1945, 33% of crops were lost to pests. In 1990, losses increased to 37% — despite a 3300% increase in the use of agricultural poisons.

With regard to controlling plant diseases, the conventional tools are fungicides and breeding. Fungicides are the quick and expensive response. In moist tropical regions, like New Guinea, potatoes are sprayed with fungicides up to 30 times per crop. This is not cheap — sometimes spraying costs more than the value of the crop. Spraying does not destroy the billions of potato blight spores that are scattered about everywhere, so the chemicals have to be used year after year. The blight fungus is certain to mutate, and become resistant to the fungicide.

Plants that are bred to be resistant to a disease eliminate the need for fungicide use. It is cheaper to plant disease-resistant crop varieties, instead of using fungicides. Also, breeding often (but not always) controls diseases for more crop cycles than fungicides do. In the long run, nature will not be controlled.

Breeding disease-resistant plants is a slow process that takes years, and sometimes the end product is not 100% effective. Breeding is not easy. It has taken scientists several years to identify the wheat genes that might provide resistance to the UG99 rust fungus, and it will take 9 to 12 years to create a finished product. This breeding process will have to be repeated for the many varieties of wheat that are being used around the world.

Six genes in wheat plants seem to provide resistance. The wheat rust is constantly mutating. It will almost certainly evolve into a form that can overcome the rust resistant breeds. This will launch a new breeding process. Eventually, the breeding process will run out of rust-resistant genes to use. Eventually, the plant breeding industry will run out of genetic tricks for holding the diseases of the world at bay.

Agribusiness corporations are engaged in a non-stop war against weeds, pests, and pathogens. Nature is constantly overcoming the defensive strategies of agribusiness, and agribusiness is constantly working to invent new (temporary) defensive strategies. Nature is not a weakling. She frequently lands solid punches on the dominators, resulting in substantial crop losses. In the long run, nature will not be controlled. The days of industrial agriculture are numbered. The era of thousand acre cornfields will not go on forever — for a number of reasons. Thousand acre cornfields are not nature's way. We need to learn this.

Breeding Wonder Crops

Green Revolution

The term *Green Revolution* describes the catastrophic population explosion that occurred between 1950 (2.5 billion) and 2000 (6 billion). A primary cause of this disaster was a series of plant breeding projects, which resulted in dramatic increases in the yields of primary crops.

OK, I lied. Green Revolution is a term that celebrates the increase in food production during this period. Our culture is so peculiar. We celebrate the brilliant successes of the brilliant scientists, and we completely disregard the horrific unintended consequences. The goal was to eliminate hunger, but the population more than doubled, resulting in even more hungry people — the tremendous success made the original problem far worse, and it created many additional new problems in the process.

The author Richard Manning has referred to the Green Revolution as the worst thing that has ever happened to the planet (a number of other thinkers have cited the development of agriculture as the winner of this award). What would the world be like today if the Green Revolution had never happened? Population-driven conflicts are prominent in the news headlines these days, and they are certain to increase in the coming years, fueled by food shortages and skyrocketing food prices.

Farm productivity in the world more than doubled between 1950 and 2000, driven by crop breeding programs. Rice and wheat were bred to have shorter stalks, so that more of the plant's energy could be devoted to seed production. Hybrid corn plants were bred to survive in greater density (more plants per acre). The productivity improvements provided by breeding were multiplied by the increased use of potent synthetic fertilizers, and the expanded use of irrigation.

The Green Revolution was all about industrial agriculture — irrigation, large farms, powerful machinery, monoculture cropping, proprietary seeds, synthetic fertilizers, pesticides, herbicides, and fungicides. Success required large inputs of water and petrochemicals. The revolution made a lot of money for a lot of corporations and fat cat land owners. It was completely unsustainable in every way, and it was a disastrous mistake. Countless small low-tech organic subsistence farms were wiped out of existence.

In recent years, crop yields have peaked, and have started to drop off a bit. The Green Revolution seems to be over. There do not seem to be new magical miracles coming down the pipeline, and vivid fantasies of a Green Revolution II are notably absent.

Meanwhile, the population explosion continues. The titans of the agricultural sector are transfixed by one big throbbing question: how can we (profitably) feed ten billion? There is substantial agreement that the one and only possible source of new crop yield miracles is genetic engineering. *But few who are close to the industry believe that this is likely.*

Please pay careful attention to the premises that this thinking is based upon:

- The problem is that there is an inadequate food supply for a growing population.
- The solution is to increase the food supply, by any means necessary.

The problem is never described as explosive, catastrophic population growth. There is little money to be made by taking prompt and effective action to reverse population growth. The brilliant leaders of our society are making almost no effort to confront this devastating horror head on. Instead, efforts are directed towards enabling the population problem to worsen. Brilliant! What could be worse for the planet than another big breakthrough in food production capability?

A third premise in this dysfunctional logic is less obvious, but provides the ancient and sacred foundation for the first two — *grow or die.*

Civilized cultures suffer from an irrational belief that perpetual growth is the primary objective of human society. Endless growth is the one and only idea in a cancer cell's mind — the survival of the host organism is not a matter for concern. Industrial agriculture does not have a spiritual connection to the generations yet-to-be-born.

Genetically Modified (Transgenic) Crops

First, we need to take a moment for semantic clarification. Plants that have their genes artificially spliced in laboratories are commonly referred to as *genetically modified* organisms (GM or GMO). This name makes purists cringe, because all ordinary domesticated plants have also been genetically modified, by centuries of selective breeding — they are no longer genetically identical to their wild ancestors. Purists insist that *transgenic* is a better word to describe organisms that have had their genes spliced. Let's try it.

Humans have been changing plant characteristics since the dawn of domestication. This was a low-tech enterprise that could be pursued by uneducated, illiterate, half-naked barbarians. It was done by manually selecting the seeds of plants having desirable traits. One hundred years ago (before transgenic science) there were thousands of varieties of domesticated soybeans, but each variety contained only soybean genes. Thus, all varieties remained within the boundaries of the soybean *genome*. In theory, all varieties of domesticated plants could have been created by nature, unassisted.

Modern transgenic engineering can disregard genome boundaries. It is capable of splicing together genes from completely different types of organisms — plants, mammals, fish, insects. Research in transgenic crop plants has been proceeding for a few decades. This research has produced some plants having herbicide tolerance, insect resistance, or disease resistance (benefits that natural mutation is almost certain to overcome). So far, it has not resulted in plants producing higher yields.

Every magnificent technology results in unintended consequences. For example, genes from cold water fish were inserted into strawberries and citrus fruit to make them more resistant to frost. Unfortunately, people with seafood allergies had reactions to these foods. In the transgenic world, it is especially difficult to predict what might go wrong.

Because transgenic engineering mixes genomes, and releases these organisms into the world, where transgenic pollen can be transported to distant regions and pollinate normal plants, the potential for serious, irreversible, and uncontrollable consequences is huge. Some transgenic plants are designed for pharmaceutical use, not food. It strains the imagination to try to envision the worst-case scenario catastrophe resulting from bioengineering, because there are endless possibilities. We are fooling around with super-powerful magic that we don't understand, and we are childishly wishing that nothing bad happens.

Transgenic plants are designed for use in state-of-the-art high-input industrial agriculture, not small organic subsistence farms. They are not intended to eliminate any of the great harms of agriculture. The corporate objective is to make farmers pay for every seed they plant. Ultimately, it's about controlling the supply of the world's food.

The wizards in their neckties smile and assure us that gene splicing is safe — trust us, we know what we're doing. It's OK. Other wizards have looked us in the eye and told us that cigarette smoking does not cause cancer. Other wizards were bursting with delight about the fabulous benefits of DDT. I'm an older man now, and I have completely lost my blind faith in the pronouncements of guys with ties who have something to sell.

Honestly, one of the worst of the worst-case scenarios for the consequences of transgenic engineering is that it became fabulously successful. Imagine a wonder plant that killed fungal spores, tasted terrible to insects, thrived in dry, salty soil, needed no fertilizers, and produced 500 pounds of grain per year. What if we became capable of feeding 20 billion people? The problem is not a shortage of food, it's too many mouths.

Junk Food

The final phase of the agricultural system is processing the foods we have grown. It is becoming quite clear that some of the foods that we eat are not good for us. They make many of us sick. They kill many of us. Every step of our industrial food-making process, from planting to the dinner plate, has serious drawbacks.

The study of food, nutrition, and health is a realm where there is little consensus, and an abundance of slick-talking hucksters. There is published research that can be used to support any dietary theory that you can imagine. Michael Pollan's books are a good place to start; he makes an effort to be fair. My favorite author on nutrition is Weston Price, because his findings and photographic documentation were just stunning.

An Epidemic of Rotten Teeth

In 1893, Weston Price became a dentist in Cleveland, Ohio. As the years passed, Price became aware of a highly unusual trend in the health of his patients' teeth — the amount of tooth decay that he observed was growing sharply — to a degree that was unknown just ten years earlier. Something strange was happening. He was watching a serious health crisis emerge right before his eyes, but he didn't understand the cause of it. Price suspected that the new problem was related to changes in his patients' diet.

His curiosity grew. Finally, he decided to do some travelling, in search of healthy people, to observe how they lived differently from us. He spent much of the 1930s visiting many lands, examining the teeth of the residents, taking photographs of them, and studying their diets. He went to remote places where people continued to live in their traditional manner — in regions including Switzerland, Ireland, Africa, Australia, New Zealand, the Arctic, and Peru. He found many people with beautiful perfect teeth, and he found many with serious dental problems, like his patients in Cleveland. Importantly, he discovered a clear difference in the diets of the two groups.

The people with happy teeth ate the traditional diet of their region, never used a toothbrush, and never saw a dentist. The people with crappy teeth ate a "modernized" diet, including white flour, refined sugar, canned vegetables, jams, and marmalades. Those who lived in remote villages in the hills and ate like their ancestors ate were fine, but those who lived by the shore and ate imported modern foods suffered for it. If one brother stayed in the hills, and the other brother moved to the city by the sea, the difference in their dental health was striking. Among those eating the modernized diet, the incidence of problems varied from place to place. In some places, only 25% of them had problems, but in other locations up to 75% were affected.

The children of those who ate modernized diets had even worse problems. In addition to tooth decay, their dental arches were deformed, so their teeth were crowded and crooked (like mine are). Their nostrils were narrower, forcing some to be mouth breathers. Their skulls formed in unusual shapes and sizes, often narrower than normal. Their hips and pelvic bones formed abnormally, making childbirth more difficult. They suffered from far higher rates of chronic and degenerative disease, including cancer, heart disease, and tuberculosis. Their overall health was often weak or sickly. Some were mentally deficient.

In Alaska, Dr. Price met Dr. Romig, who had been caring for Eskimos and Indians for 36 years. He never found cancer among those who ate the traditional diet, but those who ate trading post food commonly got cancer, and many suffered from tuberculosis, too. Romig noted that when people returned to their traditional diet, their health improved, and they got few new cavities.

Dr. Price finally went home and wrote a book to document his findings. *Nutrition and Physical Degeneration* was published in 1939. The book is loaded with his photos, and they are stunning and unforgettable. Reading it was a mind-blowing experience — an important subject that I had been ignorant of, despite 17 years of schooling.

The Roller Mill Blues

When I finished Price's book, I was disappointed that he had failed to visit a land where the traditional diet included sugar or white flour. In tropical regions, people grow sugar cane and eat it. French bread, Italian bread, and pita are traditional staples made with white flour, right?

I then began to wonder whether white flour and sugar were newer foods. As it turns out, yes, they were. In earlier days, both were expensive luxuries, and only the wealthy could afford them. The author Sidney Mintz discovered notes about a 16th century meeting — a German gentleman visited Queen Elizabeth and was deeply impressed by her black teeth when she smiled (she had a "sweet tooth"). The peasants looked much healthier than the royalty, because they couldn't afford sugar.

By the late 19th century, both sugar and white flour had become widely available and very inexpensive. The primary reason for this was because a new milling technology had emerged — steam-powered steel roller mills.

Previously, grain had been milled between stones, which ground together all parts of the wheat berry — the fiber-rich bran, the vitamin-rich germ, and the starchy white endosperm — resulting in whole wheat flour. The oils in this flour would go rancid with time, so it didn't have an extended shelf life. Another drawback was that whole wheat flour was highly nutritious, which attracted the attention of insects and other pests.

White flour was made by *bolting* — sieving whole wheat flour through fine cloth. This was a time-consuming process, so white flour was expensive, and it was produced in limited quantities. Only the rich and famous could enjoy the pleasures of white bread. The waste byproduct of the bolting process was the super-nutritious bran and germ, which was generally fed to lucky livestock.

The new steel roller mills crushed the grain, rather than finely pulverizing it. This made it much easier to separate the bran and germ from the powdered endosperm (white flour). Because of this, white flour suddenly became cheap and readily available to everyone. Since people perceived white flour to be an exotic luxury food, they eagerly consumed it in mass quantities. Because it was so low in nutrients, insects did not find white flour to be interesting at all. White flour could be shipped to the ends of the Earth, and stored for extended periods of time, without spoiling.

The wheat berry is rich with fiber, vitamins, and minerals. Most of these are stripped out when white flour is made — over 90% of the fiber is lost, and 70% to 80% of the vitamins and minerals. The bleaching process (to make the flour whiter) destroys more nutrients. Additional nutrients are destroyed by the heat generated by high-speed roller milling. Old-fashioned stone grinding was slower and cooler.

Before long, people realized that white bread was loaded with calories, but nearly free of nutrients. The government eventually passed rules requiring that white flour be "enriched" with nutrients, including thiamin, riboflavin, niacin, and iron. Despite this enrichment, white flour remains far less nutritious than whole wheat. Amazingly, 95% of the flour used in the US today is white.

With regard to sugar, the steel roller mill vastly increased production. The new mill could extract up to 85% of the juice from the cane. The previous technology could only extract 20% of the juice. So, each ton of cane could produce much more sugar, which lowered the price, and increased the output. Sugar was no longer a luxury for the rich. It became a major component of the working class diet. By 1900, 20% of the calories in the English diet were provided by sugar. Many factory workers started their day with a slice of white bread spread with sugar-packed jam, marmalade, or treacle — lots of calories, few nutrients. By 2003, the average American consumed 142 pounds of sugar per year.

Sugar consumption nearly doubled in the US between 1890 and the early 1920s — an era of rapid growth in the candy and soft drink industries. In some US cities, diabetes deaths quadrupled between 1900 and 1920. By the 1930s, the cancer rates in the US were clearly on the rise. Diabetes and cancer are far less common in societies that do not eat a Western diet.

The steel roller mill delivered a double-whammy to the world's food supply — cheap white flour and cheap sugar. These products had a long shelf life, and were shipped to every trading post in creation. Dr. Thomas L. Cleave noticed a pattern. Roughly 20 years after sugar and white flour arrived in a region, all hell broke loose. There was an explosion in dental problems, not to mention degenerative diseases like cancer, heart disease, tuberculosis, and diabetes. He named this predictable phenomenon the *Rule of 20 Years*.

Heart disease was almost unknown a century ago — except among the rich. In 1930, 1% of deaths in Britain were caused by heart disease. By the mid-1990s it was 30% and rising. The story was similar in the US. Dwight Eisenhower's doctor graduated in 1911, and at that time, he had never heard of coronary thrombosis (heart attack). By 1943, half of all deaths were caused by heart attacks. From 1907 to 1936, deaths from heart disease in the US rose by 60%, and deaths from cancer rose by 90%. But in 1953, John Gunther wrote that cancer was almost unknown among Africans.

Archeologists have provided us with a better understanding of the history of dental problems. Hunter-gatherers rarely had tooth decay. Dental problems increased just a bit with the emergence of agriculture — we ate more gummy, starchy grains, which had a tendency to stick to our teeth. Rates of decay remained quite low until about 1850, when they started rising sharply. By 1900, dental problems had become a major and serious health crisis.

Archeologists have also learned more about the history of dental health in regions where sugar cane was a part of the traditional diet. For example, on Easter Island, the rate of dental problems in skulls dating from 1700 to 1800 A.D. was extremely high (much higher than in modern Western nations). Similar data has been found on Henderson Island.

Peak Food

Over the last 50 years, the Green Revolution succeeded in dramatically increasing food production. Today, there are strong indications that the productivity gains of the Green Revolution have mostly maxed out. We are certainly approaching Peak Food — the all-time peak of food production on planet Earth. We are also moving beyond the era of cheap food. A number of factors are in play.

- Plant breeding is no longer producing varieties with dramatically increased yields.

- We are approaching the limit of plant breeding to successfully produce new varieties that are resistant to insects and diseases.

- We have reached the limit of fertilizers to increase yields (except in poor regions like Africa). In the past, cranking up the fertilizer dosage cranked up the yields, but this is no longer true. Fertilizer runoff has become a serious pollutant.

- The only conceivable source of productivity gains that would allow us to feed ten billion is transgenic engineering. *Few who are close to the industry believe that this is likely.*

- Climate change is a wild card. It has the potential to generate big challenges for agriculture.

- The end of the era of cheap energy is certain to reduce farm production.

- Growing soil salinization, erosion, desertification, and water shortages are leading us toward a period of turbulence that will make the end of petroleum look trivial. Oil is not necessary for life. Food is.

- Global grain reserves are close to the edge. In *Empires of Food*, the authors conclude: "In five of the last ten years, the world consumed more food than farms have grown, while in a sixth year we merely broken even."

How are we going to feed 10 billion? Right now, the system is breaking down, as we try to feed seven-point-something billion. We can't really even keep feeding the current population. Most new population growth will be in the developing world, where water is already in short supply.

Industrial agriculture is a dinosaur industry. Where do we go from here? Read on.

SEARCHING FOR SUSTAINABLE FOOD

In the coming decades, we're going to see big changes in the way that humans obtain food. Today, the only way to feed seven billion people is via industrial agriculture. The end of the era of cheap energy will increasingly impede industrial agriculture. Population will decline. Millions of localized food systems will come into existence. This time of change will provide precious opportunities to embark on new paths that move towards sustainability. But what is genuinely sustainable?

The civilizations of the ancient world were typically destroyed by a combination of intensive logging, over-grazing, organic farming, and perpetual growth. Their organic farming practices would likely fall within the sphere of what is today wishfully referred to as *sustainable agriculture* — organic manure fertilizers, no machines, no herbicides, no pesticides, no gene-spliced seeds.

Modern mechanized agriculture is even more destructive. For each pound of corn grown in Iowa, five to six pounds of soil are lost. For each pound of wheat grown in Washington, twenty pounds of soil are lost. Since the Europeans arrived, experts estimate that a third of US topsoil has been lost, and millions of acres have been converted to wasteland.

Civilizations have never been sustainable, and soil destruction is a common reason for their collapse. In some mountain regions, like the French Alps, a single generation observed the swift transformation from forest to field to bare bedrock wasteland.

There have been a few civilizations that have survived longer than average. For example, the Chinese have been farming the Yellow River basin for 4,000 years, and many conclude that this represents an example of sustainable agriculture. According to J. R. McNeill, most of these long-lived farming systems have a pattern. They practice an unsustainable mode of farming until chronic problems emerge, or a new technology becomes available, and then change their ways — to another unsustainable mode of farming, and then another, and then another.

History seems to imply that it is impossible to feed dense populations in a sustainable manner.

Today, you can spend many years at Harvard or Yale, get a PhD, and yet learn absolutely nothing about how the food you eat is produced. On the subject of ecological history, our education system rarely gets passing grades. While browsing essays on green subjects, it is quite common to observe the phrase *sustainable agriculture*. Does such a thing actually exist? A genuinely sustainable way of life is one that can exist *permanently* without diminishing the ecosystem.

On a whim, I spent a year hunting for evidence of genuine sustainable agriculture, because if it actually existed, then understanding how it worked could be important and beneficial. On the bright side, I did discover examples of agriculture that have the potential to be sustainable. On the downside, they are not widespread or mainstream. In the following sections, I'll share some of the things that I learned during my search.

The Nile

The agriculture along the Nile River of Egypt has been feeding humans for about 7,000 years, an unusually long duration in the history of farming. The Nile flows from south to north, and empties into the Mediterranean Sea. The Nile valley runs through the scorching hot Sahara Desert, and it receives little or no annual rainfall. Farming was enabled by annual floods, which deposited water, nutrient-rich silt, and organic matter onto the fields. Nile agriculture had many unique features:

- There was no need for farmers to add fertilizer to their fields. The floods delivered an adequate supply of nutrients in the silt deposits.
- There was no need to practice crop rotation or fallowing, because the soil fertility was not depleted by regular cropping.

- There was no need for rainfall or irrigation systems, because the floods provided adequate moisture.

- The soils did not suffer from salinization, because the floods washed the salts away.

- The crops did not suffer from the waterlogging that often affects irrigated farming, because the drainage was adequate.

- The fields did not suffer from topsoil loss. Annual silt deposits actually made the topsoil deeper.

- When croplands were expanded by creating irrigation canals, the silt deposits were not heavy enough to clog the canals (a fatal problem in other places).

Despite these unusual features, the Nile's agricultural system was not a free lunch. Every year, they received a billion tons of Ethiopian topsoil from 1,500 miles away. If we paddle upstream from Cairo, we cross the border into Sudan. At the city of Khartoum, the Nile forks into two branches. The right branch is the White Nile, which originates in Uganda. It's a slow-moving stream that doesn't send much soil to Egypt. The left branch is the Blue Nile, which originates in the mountains of the Ethiopian Plateau.

Ethiopia receives heavy monsoons in the summer months. Powerful thunderstorms pound the mountainsides with drenching rains, washing loose soils into the river. The Blue Nile is a roaring torrent during the monsoons. In Egypt, before the dams, flooding peaked between August and October, and the floodplains were covered with a thin layer of rich mud.

In 1902, the British finished the Aswan Dam in Egypt, which enabled year-round irrigation, and the production of three crops per year. On the downside, it blocked the annual deposits of fertilizing silts and humus, so farmers were required to start buying and applying fertilizers. Before long, the fields began experiencing waterlogging and salinization problems (i.e., symptoms of terminal illness). The elegant balance of the good old days was over. The wellbeing of future generations was pushed aside by the desire for increased production and profits, and the disregard for out-of-control population growth.

In 1970, the Aswan High Dam was finished. It held more water and blocked 98% of the fertile silt. It finished off what remained of the old system. Farmers were now forced to switch to modern high-input industrial agriculture. Because the location of the reservoir was very hot and dry, much of the stored water was lost to evaporation. If the dam had been built in the cooler highlands, this loss would have been reduced, but the government wanted the water to be stored within Egypt's borders.

Until the late 19th century, there was little man-caused erosion near the Blue Nile headwaters. Then came the ax, the plow, and livestock, which increased the rate of erosion. The 20th century brought rapid deforestation and desertification. In 1900, 40% of Ethiopia was tree-covered. In 2000, it was less than 3%.

Activities on the White Nile are also increasing erosion in that watershed. The old system could not have gone on indefinitely, because the Ethiopian Plateau would have eventually run out of fertile loose soil. So, even if the dams had not been built, the Nile Valley system would have eventually collapsed. The system is now on chemotherapy — petroleum, synthetic fertilizers, herbicides, and pesticides.

Today, the soil of Egypt is progressively deteriorating. Population is exploding. Egypt has become one of the world's biggest food importing nations. The Nile is an example of unsustainable farming that has survived longer than average. It is an unusual exception.

Yellow River

Civilization in China is about 4,000 years old. It began in the Yellow River watershed, where the light loess soils were easy to work with primitive digging sticks. The Yellow River flows into the Yellow Sea, and both are yellow because of the heavy load of yellow silt in the water. The nutrient-rich silt originates in the Loess Plateau. Loess is a type of soil that is highly erodible (loess is also found in the American high plains, and provided the dust for the Dust Bowl). The loess deposits in China cover about 150,000 square miles, and their average depth is about 200 feet.

Prior to civilization, the Loess Plateau was protected by forest, which helped to hold the soil in place. Then came the hard-working loggers, farmers, and herders. More and more of the precious soil was exposed, and the heavy summer cloudbursts washed more and more of it away. This initiated centuries of massive erosion, which continues to this day. In some parts of the Loess Plateau, the level of destruction is staggering and unbelievable — erosion has caused gullies 600 feet deep. Each cubic meter of water in the Yellow River contains 100 pounds of silt. Experts estimate that 2.2 billion tons of this soil are now eroded annually, twice the rate prior to agriculture.

Silt is washed into the river, which carries it briskly out of the highlands. The river then enters a vast and fairly flat plain, where the water slows down, half of the silt settles, and builds up. Because the plains do not provide a deep and durable channel for the river to flow through, the region has suffered from countless disastrous floods, which have killed literally many millions of people over the centuries. For this reason, the Yellow River has earned its nickname: China's Sorrow.

In the 16th century, some American crops were introduced to the Yellow River region, including corn, peanuts, sweet potatoes, and white potatoes. Corn and sweet potatoes did well on dryer uplands that were not suitable for rice, and were lightly settled at the time. This led to cropland expansion, which led to increased deforestation, erosion, flooding, food production, and population growth.

For centuries, the Chinese have been building dikes, in a never-ending effort to control the river, but the river stubbornly insists on freedom, and the river always wins. All flood control is temporary. Future catastrophic floods are a certainty, and the population of the basin has never been higher.

Rapid industrial growth is stimulating increased water consumption. More and more, the Yellow River is running dry before reaching the sea. The water is heavily polluted with sewage and industrial waste — one third of the water is too polluted for use by agriculture or industry. Cities are mining their groundwater.

In the Nile Valley, the river kindly watered and fertilized the fields. In the Yellow River basin, the farmers had to dig irrigation canals, use muscle power to pump the water, and continuously dredge the silt from the canals, using baskets. Dredging was required to keep the canals from being plugged with silt. The dredged silt was applied to the fields, where it provided nutrients to the soil.

China is often cited as an example of sustainable agriculture, because they have been farming the same land for 4,000 years. But, because of serious erosion problems, one-third of former cropland has been abandoned. While the farmers certainly worked extremely hard, and got bonus points for above average nutrient recycling, the longevity of this system was more due to the unique conditions in the region:

- There was a dependable source of water for irrigation.

- The soil was not salty, so they were able to irrigate for many centuries without creating salinization problems — this is the exception to the rule. More typically, the symptoms of salinization become severe within a century of irrigating.

- Like the Nile system, the river continuously delivered fertile silt to farming communities. In most other farming systems, replacing depleted nutrients is a major challenge that limits productivity.

The fertile silt is finite in quantity. Every year, the quantity of soil in the Loess Plateau is reduced by ongoing severe erosion. There will come a day when the Yellow River is no longer yellow. In recent decades, the Chinese have been shifting from subsistence farming to industrial agriculture — machines, petroleum, super seeds, and chemicals. This system is far more destructive than the preceding 40 centuries.

Slash & Burn in the New Guinea Highlands

New Guinea is an island that is close to the equator. It is much larger than the combined area of California and Oregon. The lowlands along the coast are very hot, and malaria is a major problem. The highlands have a temperate climate and fertile soils. Glaciers cover mountains that rise up to 16,500 feet high. When Australian mining prospectors entered the New Guinea highlands about 1930, they were surprised to discover maybe a million people who had previously been unknown to the outer world. People had been living in the highlands for maybe 40,000 years.

Because of the extremely rugged mountain landscape, it was nearly impossible to travel between the lowlands and the highlands without an airplane. In 1930, the highlanders were unaware of the existence of the ocean. They had virtually no contact with the people living down by the coast. Even today, there are regions of the island that are completely unexplored by the colonizers. It is believed that that there are still dozens of groups which remain uncontacted by the modern world.

The highlands were one of several places on Earth where agriculture was developed independently. Based on wee evidence, some speculate that farming may have started somewhere between 7,000 and 10,000 years ago. In fact, archeological studies are still in their early phases, and the official history of agriculture in New Guinea will not be written for years. A precise history may never be written, because 400 inches of rain per year does an excellent job of composting and erasing old evidence.

In his book *Guns, Germs, and Steel*, Jared Diamond described the limiting aspects of New Guinea agriculture. This system focused on root and tree crops. No grains or nitrogen-fixing legumes were domesticated. The primary root crops were not high in calories. The diet was also low in protein — there were no large mammals to domesticate, and they had no access to seafood. People ate spiders, frogs, mice, and people. Pigs, chickens, and dogs were imported later.

Some regions of the highlands are suitable for agriculture. Prior to the emergence of farming, these regions were covered with forest. The original farming method was *slash and burn*, also known as *swidden* or *shifting agriculture*. To make a new field, a patch of forest was cleared, burned, and planted. A number of crops were grown, but the staples were yams and taro. Fields would be used for four to six years, until the soil wore out, and then they would be abandoned for 20 to 30 years.

Yams and taro do not thrive in poor soil, so farmers preferred to clear forest for new fields, rather than till grassland. Forest soil was more fertile and friable, and it had fewer pests, pathogens, and weeds. It was also much less work to clear forest than to till grassland with primitive tools (they had no metal, wheels, or draft animals).

With time, the forest regenerated on the abandoned fields. Trees put down deep roots, which brought up minerals from the subsoil. When the regenerated forest was cleared again later, the burned wood would release the minerals to the soil. Unfortunately, each cycle of clearing and regeneration diminished the quality of the soil and ecosystem a bit. This reduced the number of plant and animal species in the forest, which diminished the diversity and vitality of the forest. Over time, abandoned fields became so depleted that forest would no longer regenerate. Forests were gradually converted into grasslands.

In his book *1491*, Charles Mann examined the slash and burn agriculture used in the Amazon. When a new field was cleared and burned, most of the nutrients in the vegetation went up in smoke — almost all of the nitrogen, and half of the potassium and phosphorus. The fires also released large amounts of carbon dioxide.

In *Out of the Earth*, Daniel Hillel described the long-term harm caused by slash and burn agriculture. For example, in Indonesia, there are more than 16 million hectares of land that is incapable of supporting either agriculture or forest. The long-extinct Mayan civilization was also based on slash and burn. Treeless Easter Island is probably the most famous slash and burn wipeout.

A tumultuous change in New Guinea highland life began about 250 years ago, with the arrival of the sweet potato, a tuber from South America. Sweet potatoes were more tolerant of poor soil than yams or taro, so fields didn't need to be abandoned so frequently. They could be grown at higher elevations, which allowed new fields to be created in unspoiled places. They produced more food — each plant could remain productive for up to two years, because new tubers kept forming.

The sweet potato soon became the dominant crop, and it was used to feed both humans and pigs. Because the food supply grew, so did the number of humans and pigs. The introduction of the sweet potato led to a population explosion. This inspired the expansion of cropland, which led to increased soil erosion. Examinations of soil sedimentation in lakes reveal a sharp increase in deposits following the shift to sweet potatoes.

When the Australians arrived in 1930, and found a million people living in the highlands, they were not observing an ancient, stable society. They were observing an unstable society, rocked by massive changes, suffering from the stresses of rapid population growth. The dense population was a recent phenomenon. From the long term perspective, then, the agriculture of the New Guinea highlands does not seem to be benign or sustainable. With the passage of time, the quality of the ecosystem was diminished.

Old-fashioned Low-impact Organic Agriculture

Some believe that low-tech horse-powered farming is sustainable. In the US, the Amish and Mennonites of Lancaster County, Pennsylvania are widely admired for their old fashioned farming techniques. They do, in fact, engage in state-of-the-art low-technology organic farming. By using horses they avoid the soil compaction caused by heavy farm machinery. They don't use chemicals on their fields, so much of the microscopic soil ecosystem remains alive, producing rich humus. This humus holds moisture very well, so that in drought summers the fields remain green and productive — even when neighboring conventional farms are toasted. They have smaller fields, surrounded by windbreaks of trees, to reduce wind erosion. The manure that they use for fertilizer is 100% organic.

Their techniques for plow-based farming are probably as low-impact as possible — but there is an impact. Here's the punch line: when these bearded farmers moved to Lancaster 250 years ago, there was 16 inches of rich topsoil — today, there is 8 inches (according to a PBS documentary). Thus, not even this often-admired system is sustainable.

In *New Roots for Agriculture*, Wes Jackson described a Kansas Mennonite's field following a heavy rainstorm. The newly-planted seeds were washed away, and the ditches were clogged with eroded black mud. Not even labor-intense, low-impact organic farming is safe from the damage caused by normal Kansas storms.

Jackson then drove to a nearby prairie, protected by a lush cover of perennial grasses, and observed that the deluge had caused no harm — in fact, the rain had invigorated the vegetation. Prairies can absorb four to fourteen times more precipitation than a tilled field, which means greatly reduced runoff. Plowed fields multiply the scale, frequency, and cost of floods.

About half of the world's remaining cropland is being degraded by erosion. The rate of erosion is related to local conditions and practices. For example, there is a belt of land in northern Europe, spanning from Ireland to Poland, where erosion is below average. In this region, the cropland has heavy soils that are not highly erodible. Rains typically come in gentle showers, rather than heavy cloudbursts. Farmland is generally flat or gently sloped, not steep. This is an unusual combination of conditions. When the farming practices of northern Europe were exported to places like Oklahoma, Pennsylvania, or Kansas, they unleashed severe erosion.

Wherever the plow is used for farming, a toll is taken on the soil, even in northern Europe. When they are not molested, forests and grasslands actually build new topsoil, and keep it in good health. Plowed soils are never in peak health.

It is important to remember that the "organic" label does not always indicate low impact, especially when organic farming is done on an industrial scale, which is a growing trend. Manure is gathered and spread by machines that consume energy. Managing weeds without herbicides requires a lot of energy — commonly by pulling a cultivator through the field several times. Sometimes flamethrowers are used to fry weeds. Insects are sometimes removed from fields by using giant vacuum cleaners. In general, industrial organic farming replaces chemical applications with increased tilling, which is injurious to the soil. It is not a win-win solution, nor is it sustainable.

At traditional old-fashioned tiny farms in China, the weeds are hand-picked, the insects are hand-picked, cultivation is done with hand tools, harvesting is done by hand, and manure and compost are manually applied. Human energy is used instead of fossil energy. This is the way of the future.

Thinkers who raise questions about organic agriculture include Michael Pollan (*The Omnivore's Dilemma*), Paul Roberts (*The End of Food*), and Stan Cox (*Sick Planet*). Organic farming eliminates nasty chemicals, but it doesn't eliminate soil destruction, a far more serious problem. Nor does it prohibit all processes that utilize petroleum products.

Agriculture has never been a highly reliable system for food production. Some years the harvest is adequate, and other years it isn't. Sometimes bad years follow bad years. Sometimes famine thins the human herd, and sometimes this is followed by further thinning from contagious disease.

In his book *The Small Isles*, Denis Rixson discusses four islands off the west coast of Scotland where the lifestyle was more primitive than average. Herding and fishing were primary occupations, and a small amount of wheat was grown to provide winter bread. The people were healthy and long-lived. One old gent on the Isle of Rum is mentioned in a 1764 report. He died at the age of 103. He was 50 before he first tasted bread. He criticized the younger men for raising wheat, because he "judged it unmanly in them to toil like Slaves with their Spades, for the Production of such an unnecessary Piece of Luxury."

The peasants of Rum were fortunate to have access to fish and grazing land. Most of the other peasants of Europe were forced to toil like slaves. In a similar vein, the Maasai herders of Kenya consider farming to be a repulsive activity that is beneath their dignity. They will not even dig the soil to bury their dead. Dying people are carried out of the dwelling and left for the vultures, hyenas, and other wild animals to enjoy.

In *Akenfield*, Ronald Blythe interviewed elderly English villagers who still remembered 19th century life. Today, farmers wear out machines, but in the old days farming wore out the farmers' bodies. Farming sucked every ounce of strength out of them. All of the villagers had experienced being really hungry. Many did not even own a change of clothes.

In his book *The Little Ice Age*, Brian Fagan describes what life was like in the age of muscle-powered agriculture. In his words, it was "unrelentingly harsh." It was brutally hard work — in excavations of old cemeteries, archeologists have discovered that spinal deformations in peasant farmers were common. Most adults suffered from arthritis.

Contagious disease was common, malnutrition was common, and the average lifespan was brief. Even in good years, most peasants lived on grains — bread, porridge, gruel. If they were lucky, they tasted some meat or fish at a Christmas feast. The nobility lived fairly well, but the vast majority of agrarian society enjoyed poor diets, short lives, and hard labor.

Flooded Paddy Rice Farming

Rice is an interesting crop. Rice is second only to corn in world grain production today. Most corn is fed to animals, factories, or automobiles (as ethanol). Most rice is fed to humans, and most of it is eaten close to where it was grown. Rice is the staple food for 60% of humankind, the source of one-third of our calories.

Rice can be grown on dry land or in flooded paddies. Dry land farming presents the customary drawbacks of till-based agriculture. About half of rice farming is done on dry land. On average, dry fields yield one-quarter to one-third as much rice as flooded paddies.

Flooded paddies minimize erosion risks, prohibit the growth of most weeds, and deter rodents and bugs. The wet environment makes water-soluble nutrients readily available for plants. A major benefit is that water ferns and algae fix nitrogen, making this vital nutrient available to the plants throughout their life cycle — this is in addition to the nitrogen provided by the manure and sewage added to the fields (or the application of modern chemical fertilizers). This nitrogen bonus, combined with abundant moisture, makes paddy farming much more productive than dry land farming.

As techniques improved, and flooded paddy farming expanded, much more food was produced, which led to the births of many more rice-eaters. Population pressure has forced the expansion of rice fields up the sides of steep hills, where crops are grown on terraced plots. This presents greater erosion risks than flat lowland fields, especially when the maintenance of terraces is neglected.

It takes large quantities of fresh water to grow rice in flooded paddies. In some regions, this water comes from groundwater mining, which is not sustainable. In California, where rice is grown on the desert, water comes from heavily-subsidized water projects that siphon water away from other regions. The declining availability of fresh water is going to put sharp limits on the expansion of rice farming in the future. Some worry that rising sea levels are going to create big problems for rice growers in the coming years, because much rice is grown in low elevation wetlands.

Methane is a greenhouse gas that is 20 to 25 times more potent than carbon dioxide. Flooded rice paddies generate substantial methane. Methane production is increased by adding manure, crop residues, compost, or chemical fertilizers. Flooded paddy rice farming generates about 30% of global methane emissions, and is likely the number one source of human-generated methane emissions. Thus, increasing rice production to feed a growing population will also increase methane pollution and climate change.

Flooded paddy rice farming fueled population growth. At the same time, it increased the death rate by providing a breeding ground for diseases like malaria, schistosomiasis, filariasis, arboviruses and other vector-borne diseases. A hundred years ago, 90% of Chinese had worms, and 25% died from fecal-borne infections. No amazing technological advance is free.

Malaria is spread by mosquitoes, and mosquitoes thrive in stagnant water. In some regions, the name for malaria was *rice malaria*. For many years, it was against the law to create rice fields near Spanish or Italian cities. In 1864, long before the era of insecticides, the venerable George Perkins Marsh wrote: "The cultivation of rice is so prejudicial to health everywhere that nothing but the necessities of a dense population can justify the sacrifice of life costs in countries where it is pursued."

Insecticides have had mixed results.

- Spraying in Guyana killed the mosquitoes, which reduced the death rate, which spurred population growth, which forced an expansion in rice farming.

- Insecticides commonly kill mosquito-eating creatures, like frogs and birds, and this is often followed by having more mosquitoes than there were prior to spraying.

- Mosquitoes develop resistance to insecticides, and the cost of spraying is increasing.

- Some efforts have been successful in eliminating malaria from a region. But an infected person cruising through the region can get bit by a mosquito and re-infect the region.

As we move beyond the era of cheap energy, the use of farm chemicals will fade away, much to the delight of the mosquitoes, frogs, and birds. On the other hand, the area of flooded paddy farming is likely to shrink, as it becomes too expensive to run motorized irrigation pumps, and as the rising seas submerge coastal wetlands.

In his book *The End of Food*, Paul Roberts adds another dimension to the health issues related to rice paddies. Viral diseases, like influenza, can readily mutate and trigger deadly pandemics, like the one in 1918 which spread around the world in two months, killing up to 100 million. Humans are vulnerable to a number of viruses, and so are other animals. Usually, poultry catch poultry viruses, and humans catch human viruses, but once in a while humans might catch a mutated poultry or swine virus. This is what happened in 1918. Humans acquired a lethal influenza that suddenly mutated and became highly contagious, too. My grandmother's sister, Emma Amundson, died of the flu on 19 November, 1918. She was a beautiful young nurse.

Everyone expects that epidemics of contagious diseases will periodically appear in the coming years. We live in a highly mobile world, which is extremely overpopulated, where many millions are malnourished, and have weakened immune systems. It's a paradise for disease.

If we had to guess where the next deadly pandemic might originate, China would be near the top of the list. China has dense populations of humans, ducks, chickens, fish, and swine living in close proximity. If influenza viruses are going to jump from species to species, China provides ideal conditions. For example, migratory wild birds may carry an avian influenza virus from another region and deposit it in a rice paddy. The farm ducks pick it up, and pass it to the chickens.

The air in the chicken coop may be filled with virus particles, and the farmer might inhale them and develop an infection. The infected farmer might transmit the influenza to the rest of humankind.

Following the 1918 pandemic, the Coast Guard discovered several isolated Eskimo villages in Alaska where everyone had died. These people had no contact with outsiders; they had been infected by migrating birds. Today, wild bird flocks that are known to be infected with influenza are tracked by satellites. Still, outbreaks will not be easy to control, because modern society is so highly mobile. Public health experts shudder whenever a new strain of flu virus appears — and they appear every few years.

The Green Revolution dramatically increased rice yields. It combined improved plant breeds, generous applications of synthetic fertilizers, large inputs of fresh water, and lots of insecticides, herbicides, and molluscicides. Pesticide poisoning is common among rice farmers — the toxins accumulate in their bodies.

Sharp increases in population are projected for the rice-eating regions of the world. Many people in these regions are undernourished today. Conventional thinking concludes that the solution to this problem is to increase food production. Experts in rice growing are not bursting with bright ideas about how to achieve this. Plant breeding has produced few significant advances since 1965. Adding more fertilizer no longer increases yields.

The current population is kept alive because of advances made during the Green Revolution. Old-fashioned low-tech farming would no longer be able to feed the current numbers. Yet the end of the era of cheap energy is going to force a return to old-fashioned low-tech farming. Obviously, the most densely populated region of the world is moving directly toward a monumental food crisis.

The Three Sisters in the Mississippian Era

The three sisters are corn, beans, and squash, and they were domesticated by Native Americans who lived south of the US border. Judging from old written records and drawings, the three sisters intercropping method was very common, and was likely the dominant mode of corn-based agriculture. The three sisters planting system was used in many regions of North America and South America.

The Indians of the New World had no draft animals, plows, wheels, or metal tools. Given these limitations, the three sisters system was a clever low-tech process that produced high yields per acre — often far more than the farmers of Europe. It also had some serious ecological drawbacks.

The three sisters cropping system is frequently referred to as sustainable agriculture. Was it? In this section, I'm going focus attention on how the system worked in the eastern half of the future United States, because this region has been studied and documented more thoroughly.

The three sisters were planted on mounds of soil that were typically spaced about three feet apart. The soil was scraped into mounds by using hoes; the fields were not tilled. Several corn seeds were planted near the top of the mound, and beans and squash were planted around them. The beans used the corn stalks as poles to climb on, and the squash spread out across the ground. The large squash leaves shaded out weeds, and reduced evaporation. The spiny squash vines discouraged the activities of small animals.

Beans are legumes, and they helped to maintain nitrogen levels in the soil, a crucial service. Dried beans can be stored long-term, and they are a good source of protein. Squash seeds are high in protein, and rich in oil. Corn is a calorie-packed super-food. Most of the world's early civilizations were centered around growing super-foods — corn, wheat, rice, and potatoes. Corn fueled the rise of civilizations like the Aztecs, Mayans, Toltecs, Olmecs, and Anasazi.

Corn farming trickled northwards, into the future US, for several centuries prior to 800 AD. Corn was not a mainstay food at this time. But around 800, corn-farming shifted into high gear, and rapidly spread and intensified. This stimulated rapid population growth, and the development of many cities and towns. Beginning in the southwestern states, serious corn-growing expanded along the basins of the Mississippi, Missouri, Illinois, Tennessee, and Ohio rivers. The corn culture reached Iroquois country in New York by 1350, and finally spread up into southern Ontario.

Eventually, most of the region east of the Mississippi River was inhabited by corn growers who also hunted, fished, and foraged for wild foods. When the Europeans arrived, a significant portion of the future US was a corn-based agricultural society. Almost no Indian farming was done in California, the Pacific Northwest, the Great Basin, and most of the plains region. In his book *Indians of North America*, Harold E. Driver estimated that less than half of North America was inhabited by farmers, but 90% to 95% of Native Americans ate crop foods, indicating that farm country was densely populated.

The new corn-fueled era is known as the *Mississippian culture* (800 AD to 1500 AD). Its four biggest cities were Cahokia, Illinois; Spiro, Oklahoma; Moundville, Alabama; and Etowah, Georgia. Each city built mounds and other monumental earthworks, manually moving millions of tons of dirt using baskets.

The city of Cahokia had up to 30,000 residents at its peak (950-1250). Wood was the fuel used for heating and cooking. Having no axes or wagons, acquiring fuel wood would have been a major job. Eventually, the daily effort needed to gather fuel for 30,000 could have been so great that the existence of the city was no longer practical. Perhaps this was the first North American energy crisis. The large population also took a big toll on the number of deer, buffalo, and other game in the region (butchered dog bones have been found in great quantities). Other possible reasons for Cahokia's collapse include soil degradation and a cooler climate.

The Indian corn-eaters commonly practiced slash and burn agriculture. Forest would be cleared, and a field would be farmed for several years, until the yield declined. Then a new field was cleared. After lying fallow for some years, an abandoned field could recover some fertility, and be used again, but fields could not be recycled indefinitely. Eventually, after 100 years or so, the community would be surrounded by worn out fields, and have to move to a fresh forest, build a new homes, clear new fields, and start over again. Moving provided access to fertile soil, and a fresh wood supply — and it left behind pests, weeds, invading grasses, badly depleted soil, and the "evil spirits" that caused the decline.

American history tends to present Indians as primitive savages. When I graduated from college in 1974, my conception of native agriculture was that it consisted of a few big gardens, widely scattered across a vast forest of unspoiled wilderness. Indians were primarily hunters, for whom gardening was an enjoyable hobby. Reality is rather different. In many regions, farming provided most of the calories. It is fascinating to read the observations of the early European explorers and commentators:

- In 1540, Hernando de Soto visited many towns in Florida. This region was populous and thriving at the time, before smallpox exploded. His saga described walking four leagues (about 12 miles) through cultivated land at one location.

- In 1669, a commentator noted that Haudenosaunee villages were often surrounded by six square miles of fields, according to Charles Mann.

- In the upper Ohio River valley and New York, some cultivated areas covered hundreds of acres, according to Paul Weatherwax.

- In 1779, an Iroquois village near Rochester, New York had fields extending for several miles, according to Michael Williams.

- Benjamin Hawkins travelled through Georgia in 1798 and 1799. He described one Indian farming location that followed a river for four miles, and was 100 to 200 yards wide. He saw another site where there were 1,000 acres of abandoned fields.

The picture that emerges is that Indian farming was much larger in scale than mere hobby gardening. Most farming was done near the shores of lakes, rivers, and streams, because this is where most people lived. Many forests remained. The rich soils of the prairies were not farmed, because breaking up the heavy sod was too much work. It was much easier to clear forest.

Even before modern hybrid seeds and synthetic fertilizers, corn was an amazingly productive crop. Commonly, each seed planted yielded 300 at harvest time (wheat was more like 30 to 50). Not coincidentally, corn is a "heavy feeder," with regard to soil nutrients. Indians very carefully selected locations for new fields, and chose only those places having the finest of soils, generally near rivers. Corn quickly wore out even the richest, most fertile soils.

While the beans replaced nitrogen, little was done to replace other mineral nutrients, like phosphorus or potassium, that were carried out of the field in the form of corn, beans, and squash. The North American Indians had no manure-producing livestock. In New England, there were massive herring runs in many streams. Indians would plant one or two fish in each mound. William Cronon wrote that in this region, fields could be used for ten or twelve years. South of New England, observers reported that the Indians used no fertilizers — on a new field, the first harvest was great, the second was good, and the third was marginal.

The process of clearing a new field included burning off the brush. Fields were also burned annually, prior to planting. Paul Weatherwax reported that this burning damaged the humus of the soil, and the remaining humus was soon lost to leaching and erosion. Humus is an important component of healthy soil.

Allowing depleted fields to return to forest was not like plugging them into a battery charger, and restoring them to full fertility in a decade or more. Over time, the forest soil recharger would add nitrogen to the soil, and replace the destroyed humus. The tree roots would extract nutrients from deeper soils, and deposit these nutrients on the surface, as their leaves, seeds, and woody parts fell to the ground. Recharging was a slow process, and its effectiveness is unclear. Obviously, the supply of mineral nutrients in the subsoil was finite, and could not support an extractive system indefinitely.

In his book *1491*, Charles Mann discussed the soil erosion that always goes hand-in-hand with forest clearing. Archaeologists have been doing soil sedimentation studies, and two things stand out. Erosion increased sharply after the corn culture moved into a region, and, at the same time, pollen from bottomland trees almost disappeared. There was a lot of erosion and flooding between 1100 and 1300, from the Hudson Valley to Florida. Many corn fields were located in or near flood plains.

Scholars have divided the Mississippian culture into three eras:

- In the Early Mississippian period (800-1100), corn became a primary food, the practice of corn-growing spread widely, and there is little evidence of high level warfare.

- In the Middle Mississippian period (1100-1350), the big cities were built, population grew, and this generated a lot of social friction.

- In the Late Mississippian period (1350-1450), the culture declined, and the big cities were abandoned. There is evidence of a cooler climate worldwide (the Little Ice Age), and several extended dry periods.

During the middle period, when population grew, there was increased competition for fish and game. Many cities and towns were surrounded by heavy wooden palisades, indicating significant levels of warfare. Destroying corn fields and stored grain was a primary military objective. As a result of social turbulence, a number of tribes migrated from region to region — some advancing, some fleeing. Many settlements were built, and many were abandoned. Corn country was not a stable, peaceful utopia.

At the same time, social structure was shifting from the egalitarianism of nomads to the hierarchical class system of agricultural society. In the big cities, there was a powerful ruling elite. One high status man at Cahokia was buried with 53 young women, who were neatly arranged in rows. In other areas, there was an increase in chiefdoms, where everyone was no longer equal. Power became more concentrated, and so did wealth, in the form of stored corn. (This pattern of agriculture, conflict, chiefdoms, and palisades is quite similar to the pattern of early farming societies in many regions around the world.)

Scholars speculate that many mound sites mark the location of chieftain headquarters. Around 1540, the saga of de Soto describes powerful chiefs who ruled over domains spanning hundreds of miles. In Georgia, the city of Coosa was the headquarters of a *paramount chief*, who ruled over several smaller chiefdoms.

Persistent warfare inspired the formation of regional alliances for mutual defense, like the League of Five Nations in Iroquois country, and the Four Nations confederacy in Muscogee country to the south. Were these alliances the embryonic forms of new civilizations? It is fascinating to contemplate what America would look like today if it had not been discovered.

Because of the large number of completely unrelated language families that existed in the eastern US, it is unlikely that the corn culture was spread by the military expansion of an empire based in Mexico. More likely, the corn culture spread by adaptation, from group to group, like the bow and arrow did. When a neighboring tribe became farmers, and their population swelled, the future of the traditional hunter gatherers in the region was doomed.

In the US, the Indian corn culture was crushed during its growth phase by the arrival of Europeans and their diseases. It was not allowed to continue on its original trajectory. South of the border, where the corn culture began, it did not stabilize and mature into a sustainable way of life. Mexico has not been an ecological paradise for a long, long time.

It doesn't appear that the three sisters cropping system is an example of sustainable agriculture. These farmers were far harder on the land than the tribes of nomadic foragers living to the west — but they caused much less harm than modern industrial agriculture does.

Paleopathology is the study of diseases in ancient eras. Paleopathologists have studied the skeletons of thousands of Native Americans found in the Illinois and Ohio River valleys. They discovered remarkable differences between the earlier nomadic foragers and the later farming people. The skeletons of nomadic foragers were boring to study, because most of them revealed wonderfully healthy people. In *The Ecological Indian*, Shepard Krech wrote that it was not uncommon for Indians to consume six to twelve pounds of meat per day (the Hudson Bay Company's ration was seven to eight pounds of meat per day).

Following the shift to corn eating, the skeletons became more interesting to study. There was a big increase in dental problems, anemia, osteoarthritis, syphilis, and other degenerative diseases. Tuberculosis swelled to epidemic proportions. One-fifth of newborns died before the age of five.

Aquaculture

Seafood is yet another realm where our wild ancestors had the most intelligent, efficient, and sustainable approach. They kept their communities small, they let the wild fish raise themselves in perfect freedom, and they only took what they needed. Their ingenious system required no money, no ships, no petroleum, no freezers, no hatcheries, no canneries, no fish farms, and no employees. Fishing wasn't a full-time job. It was fun, exciting, nourishing, and delicious.

In ages past, most of our ancestors lived near rivers, lakes, and coasts, because that was where the food was. Fish, shellfish, waterfowl, and marine mammals lived in healthy abundance in many places until the 19th century. It was nutritious food, and it was perfectly sustainable as long as the human population remained modest, and the fishing, hunting, and gathering did not mutate into commercial enterprises that over-harvested.

Unfortunately, industrial fishing has largely replaced subsistence fishing. The fisheries of the world have been heavily depleted in the last 200 years, driven by two simultaneous disasters: a population explosion and the Industrial Revolution. Many new technologies have increased the seafood industry's ability to harvest and sell ever-increasing quantities of wild fish. But the living creatures of the sea are finite in number, thus continuous growth in seafood production is impossible. Most of the oceanic fisheries have collapsed or are approaching collapse.

The fish mining industry strongly resists rules and regulations, ignores the obvious warning signs, and typically continues fishing until the nets come up empty. Brilliant! Enforcement of fishing laws is generally weak or non-existent. In many places, a substantial portion of the fish brought to market were caught illegally. The fish mining industry is destroying itself.

Another serious problem is that aquatic ecosystems sometimes fail to recover from overfishing. For example, the Newfoundland cod fishing industry was shut down in 1992, to allow the depleted fish stocks to recover. They still haven't recovered. Some believe that this is because there are far fewer small fish for the cod to eat. The fishing industry is catching massive numbers of small fish, which are used to make feed for hogs, chicken, pets, and fish farms.

While industrial fishing is wheezing, the human population explosion continues, and the demand for seafood is growing. This is spurring rapid growth in the aquaculture industry. Most of modern high-tech saltwater aquaculture is run by corporate interests. The big players are not interested in providing affordable protein for low income people. The big players are interested in maximizing profits by selling high-priced seafood to the well-fed elite. They are not interested in taking excellent care of the aquatic ecosystems where they operate. The wellbeing of the generations yet-to-be-born is not a corporate objective.

Salmon and shrimp are the dominant products, and they are raised in high density on farms located along saltwater coastlines. The overcrowding encourages diseases, pests, and pollution. Paul Molyneaux's book, *Swimming in Circles*, shines a bright light on the dark side of saltwater fish farming.

Shrimp farms tend to be very profitable in the short term, but are so vulnerable to viral diseases that few of them have a long term. The viruses readily spread to neighboring shrimp farms, and out to wild shrimp. The construction of new shrimp farms often results in the destruction of valuable mangrove ecosystems, which is a great tragedy.

Farmed salmon suffer from bacterial diseases like salmonicida and furunculosis. So their feed is laced with antibiotics, which kill weak pathogens, encourage the evolution of drug-resistant pathogens, and make the salmon grow bigger. Salmon are also vulnerable to killer viruses, like infectious salmon anemia. Diseases readily spread from farm to farm, and also to the wild salmon living in open waters.

Dense populations of farm salmon attract dense populations of pests called sea lice. In waters near Scottish fish farms, the lice have driven away the indigenous wild trout. To control the lice, the farmers use pesticides. Thus, the salmon meat contains pesticide residues, and so do the wild scallops that live in the vicinity. The pesticides can kill nearby lobsters.

The outgoing tides flush the salmon farms' toilets, carrying excrement, feed waste, antibiotics, pesticides, and lice out into the open sea, for the enjoyment of the other sea life. Nitrates in the pollution are converted into ammonia, much to the delight of toxic bacteria — amnesiac shellfish poisoning shut down 10,000 square miles of scallop beds off western Scotland.

About one-sixth of the wild fish caught by the fishing industry are turned into feed for aquaculture, chicken, hogs, and pets. They are being caught at an unsustainable rate. The warning lights are flashing. This diminishing resource will limit the future of aquaculture based on raising carnivorous saltwater creatures.

Two to five pounds of perfectly edible small fish, like sardines, mackerel, herring, or whiting are used to create each pound of farm salmon — high-profit luxury food for poorly-informed rich people (who also love shrimp and tuna sushi). Living lower on the food chain, these smaller fish accumulate a lower toxin load, so they are less harmful to eat than salmon. We could just eat the little fish, and skip the farms and their many problems, but this would feed more people, be less profitable, waste less energy, and be better for the environment. (Gasp!) Better yet, we could sharply reduce our population, catch fewer fish of all sizes, and allow the aquatic ecosystems to heal.

Unfortunately, the small wild fish are not free from contaminants. Fish living in the vicinity of industrial areas accumulate higher levels of toxins in their fatty tissues. When a salmon eats a smaller fish (directly, or in feed), it also consumes its toxin load. Some toxins are not excreted, and build up in the salmon (bioaccumulation).

Most of a wild salmon's diet consists of zooplankton, krill, and small fish. A farmed salmon's diet majors in fish feed, which includes fish meal and fish oil. It is a high-powered food intended to promote rapid growth. It also contains pesticides (for sea lice), antibiotics (for disease prevention and increased growth), and dye (to make the meat orange, imitating the appearance of healthy wild salmon).

Because of their natural diet, wild salmon consume and accumulate fewer toxins. One independent study, funded by the Pew Charitable Trusts, tested 700 wild and farmed salmon from around the world. It found that farmed salmon, on average, contained ten times more PCBs, dioxins, and pesticides than wild salmon.

The salmon farming industry assures us that farmed salmon is good for us. They are experimenting with less toxic fish feeds. But a number of health experts warn that young children, women of childbearing age, pregnant women, or nursing mothers should avoid eating farmed salmon. Under growing pressure from consumers, Norwegian salmon farmers have reduced their antibiotic use by 90%.

Luckily, the highly unwholesome farming of salmon and shrimp produces less than 10% of the world's aquaculture output. At the opposite end of the aquaculture spectrum is old-fashioned, low-impact, low-budget, freshwater aquaculture. The Chinese have been fish farming for 3,000 years, and their primary products include carp and tilapia — vegetarian fish that do not require an expensive diet of fish flesh. Chinese fish ponds were family-based enterprises designed to convert "wastes" into low-cost high-quality protein, and high quality fertilizer. Inputs included rice bran, leaves, grass, and fecal material from livestock, poultry, and humans. It was common for latrines to be built directly over fish ponds. Hog and poultry facilities were also built over fish ponds.

Fish ponds are often smelly and covered with green slime and aquatic plants. The rich nutrients in the water promote the growth of algae, phytoplankton, and aquatic plants, which provide nourishment for the fish. Plants that grow in the ponds, like duckweed and the Azolla fern, are food for chickens and ducks. Crustaceans and snails from the pond are fed to poultry. Some clever Chinese farms raise a polyculture of four types of carp, each utilizing a different segment of the food chain — phytoplankton, zooplankton, vegetation, and detritus.

China is the world leader in aquaculture. This freshwater industry is rapidly growing, and it has great potential for future growth. Modern freshwater aquaculture uses grain-based feeds, to increase productivity. Tilapia is emerging as the preferred fish for fresh-water aquaculture. They grow very quickly, they are less boney than carp, and they breed abundantly, without human assistance. They convert feed into meat more efficiently than cattle, chickens, or hogs. They can survive better than other fish in water having high ammonia levels from urine inputs. In some places where they have been introduced into open waters, they are becoming an invasive species nuisance.

In Kolkata, India (formerly Calcutta), carp, tilapia, and catfish are farmed in ponds at the sewage treatment plant — the biggest sewage-fed aquaculture facility in the world. The primary source of nutrients is sewage, no fish feed is used. This is the norm in Indian aquaculture. Productivity could be increased by feeding fish cereal products, but cereal feed costs money and sewage is free. The fish grow to marketable size in about four months. The farmers carefully monitor water quality, so that the fish are not suffocated by oxygen depletion. When the fish have reached market size, the pond is drained, and the nutrient-rich silt is gathered up and spread on nearby fields. This enterprise annually converts sewage into 18,000 tons of affordable high-quality fish.

In Vietnam, they farm a type of catfish (tra) that can survive in stagnant latrine ponds having very low oxygen levels. It is capable of coming to the surface, opening its mouth, and breathing air. It is also capable of surviving high-density stocking.

Raising fish in ponds with fecal inputs has some public health concerns. Ponds that mingle the virus-laden feces of humans, poultry, waterfowl, and/or swine can become breeding grounds for new variants of influenza that could be highly lethal and highly contagious. If migratory birds visit the pond, and acquire a new and dangerous virus, they can spread it to distant lands, bypassing quarantine controls.

Fish farm workers have elevated risks of getting parasites like tapeworms, roundworms, and hookworms. Cholera, typhoid, and other pathogens can survive in the ponds for a while, so workers should have appropriate training and protective clothing. With proper handling, and thorough cooking, the pathogens are not transmitted to consumers.

Turning free, locally produced sewage into protein is clever, and possibly sustainable. It eliminates the need for grain-based feeds, and the many problems caused by growing grain. It doesn't require fish-based feeds, which take a big toll on wild fisheries. It doesn't require huge ships, and large quantities of petroleum. It elegantly eliminates the need for energy-guzzling, sludge-generating sewage treatment plants. Freshwater farm fish are raised in ponds, out of contact with wild fish. The ponds are not interconnected, so outbreaks of diseases and pests are kept isolated. Of course, only pure, precious, wholesome sewage should be used — not sewage containing industrial wastes, household chemicals, or other toxic pollutants.

Pond-based aquaculture can be an important component in the transition to a sustainable future. The unfolding collapse will reduce the human population, terminate unsustainable industrial fish mining, and eliminate many dead zones. Dams will eventually fail, re-opening vast areas of fish habitat. Once wild fish recover in numbers, and the human population returns to healthy levels, there will be no need for aquaculture.

Unfortunately, the primary trend in freshwater aquaculture today is toward industrial-scale production, utilizing high-potency commercial feeds, antibiotics, intensive management, and energy-guzzling technology.

Grass-fed Grazing

Our wild ancestors were not interested in enslaving animals and becoming herders or ranchers. They let wild animals raise themselves, because it required far less work, and the animals really did a superb job of taking care of themselves, despite their lack of human managers.

Our wild ancestors sometimes deliberately set fires to keep prairies open, to provide optimal grazing conditions, and to eliminate cover for the lions and tigers and bears who enjoyed having humans for lunch. To some degree, this burning was simply an imitation of the lightning fires that nature lit every year. (Formerly, the mammoths and mastodons had limited the spread of woody vegetation in the Americas.)

In a brilliant conflict-avoidance strategy, nobody owned a wild buffalo, elk, antelope, or deer. Nobody owned the prairie. Nobody owned the fish. Nobody owned the birds. This was an era when people lived in Nature's way, and Nature was a fierce believer in absolute freedom for all. It was an era of wolves, not dogs — and human beings, not consumers.

When the Europeans came to America, they brought with them a system that was the exact opposite of Nature's way. It was an elaborate system of ownership, control, exploitation, with a hunger for wealth and status — a brilliant strategy for encouraging perpetual conflict.

Surveyors spread out across the wilderness, dividing the land into counties, townships, and sections. The wild and free were killed or driven away. The land was sold off, logged off, fenced off, plowed up, and taken over by domesticated plants and animals. In regions that provided ideal conditions for farming, wildlife habitat — and the wildlife — were essentially eliminated. Almost nothing remains of the Corn Belt's vast original prairies and their happy wild inhabitants.

Thus, it's not easy to defend veggie burgers as an ethical means for reducing cruelty to animals. Wandering through every soybean field are legions of the furious ghosts of mass-murdered buffalo. Year after year, the displaced animals are never allowed to return to their homeland. Few have survived. Their wild homeland has been transformed into an unnatural synthetic ecosystem. Every time a farm machine moves through a field, healthy wild animals are killed.

You also have to wonder if it is ethical to eat vegetarian foods that come from organic farms that engage in soil mining. Every loaf of bread carries a cost in lost soil. Good food is hard to find — nuts, berries, plums, dandelions — and even these can be produced in a harmful manner. *The Vegetarian Myth* by Lierre Keith carefully explores the realm of food, and its many impacts.

Anyway, in the new European paradigm, the trees were worth money, the fish were worth money, the skins of animals were worth money — everything that could be converted into property was worth money. A person's social status was not based upon their honor, courage, honesty, generosity. Status was based on wealth. The way to prove your worthiness, grow in status, and receive praise and admiration from society was to cut more trees, catch more fish, trap more beavers, mine more ore, and so on. This was not a path of balance. The European paradigm cherished wealth, not health.

When the Europeans arrived, large herds of buffalo were common throughout the Midwest, and as far east as New York, Pennsylvania, Georgia, and the Carolinas. They ranged from Mexico to Great Slave Lake in Canada. The buffalo required no corn, wheat, or soybean meal, so they had no need for farms, farmers, fences, roads, railroads, or ranchers. The existence of the herd was not a cause of massive soil erosion, large dead zones, or groundwater mining for irrigation. Indeed, the herd lived in exquisite balance with the prairie ecosystem, according to Richard Manning, in his outstanding book *Grassland*.

Like buffalo, cattle do not need to be fed grain. Grain is hard for cattle to digest. It often makes them sick, and sometimes kills them. In fact, the vast majority of the world's cattle never see a feedlot. Most cattle spend their entire lives in the beautiful great outdoors, eating the food that they find to be the most delicious and satisfying — grass.

Following World War II, the Green Revolution stimulated the production of massive surpluses of corn, which led to the expansion of feedlots. The US government pays generous subsidies to farmers to grow lots of grain. *Somewhere between 60% and 80% of the grain we raise is fed to livestock!* Typically, a US calf will spend its first six months eating grass on a ranch. Then it will be sent to a filthy feed lot for *finishing*, where it will spend its last 150 days fattening up on antibiotic-laced feeds made from soy, grain, and poultry litter (shit & sweepings). About 78% of US cattle end up in feed lots for finishing.

The amount of land that it takes to grow the grain needed to finish a feedlot steer is about the same as the amount of good pasture it would take to finish the steer on grass. But finishing with grain is faster, and the grain business is far more lucrative for the agribusiness corporations that sell machinery, fertilizers, seeds, chemicals, and so on. According to Paul Roberts, in a CAFO, a six month old 500 pound feeder calf can be turned into a 1,350 pound steer in four months. Grass-fed cattle take two years to reach 1,100 pounds. Don't we have time for better food, healthier soil, and happier cattle?

The soil destruction and nitrogen pollution caused by growing grain and soy are two huge and very serious problems, and they could be substantially reduced by simply allowing cattle to eat their preferred natural food — grass. The feedlots could be shut down. We could cease our senseless war on the soils of the Midwest, allow the prairies to return, and enjoy healthier food and a healthier ecosystem. We could turn the Corn Belt into the Buffalo Belt.

Research at the Land Institute concluded that well-managed pasture can produce more carbohydrates and proteins than a cornfield. In a pasture, the animals consume a large portion of the vegetation that grows there, and convert the cellulose into protein. In a cornfield, all that's retrieved are the kernels. Traditional dairy farms convert grass into milk products and meat, producing four times more calories than raising animals for meat only.

In his book, *Altars of Unhewn Stone*, Wes Jackson describes a 6,400 acre prairie ranch in the Flint Hills of Kansas that has never been plowed. It supports 1,700 cattle during the grazing season, and it is mostly managed by one cowboy. There is no soil erosion, and no fertilizer is added. Energy use is mostly related to the cowboy's pickup truck. The operation is almost entirely sun-powered. It produces nutritious, organic food in a manner that is sustainable (minus the truck). In Jackson's opinion, even the most reckless rancher causes less harm than tilling (he hasn't visited the new deserts in Mongolia, where goats that produce cashmere wool have become hoofed locusts).

Most of the meat for sale in America today is dubious stuff, produced in an efficient, profitable, unnatural, and nasty manner. Many nutritionists warn us that meat is unhealthy, and can even give us cancer. In the 1930s, Dr. Weston Price travelled the world, in search of healthy people, to see what they ate and how they lived. In Kenya, he visited the Maasai and the Muhima, two tribes of herding people who ate mostly blood, milk, and meat. They were radiantly healthy. The doctors that Price met in Africa all indicated that cancer was almost unknown on that continent. About 12% of humankind survives on a diet based almost entirely on foods obtained from ruminant grazing animals.

When John Gunther toured Africa in 1952, the doctors he met also reported a near absence of cancer. Dr. Price was extremely suspicious about the health effects of sugar and white flour, not animal foods. Of course, modern animal foods are produced in a manner completely unlike traditional Maasai herding.

Some people believe that the livestock industry is a primary cause of climate change. Two hundred years ago, when much of America was still Indian country, there were 50 to 100 million buffalo, 10 million elk, 3 million horses, and many millions of deer and antelope. Today, most of these wild grazing animals have been replaced by 97 million cattle. The numbers of large grazing animals did not skyrocket, they may have declined. What did skyrocket was the population of extremely wasteful, pollution-generating humans. Large grazing mammals are an essential, necessary component of healthy grassland ecosystems.

With regard to greenhouse gases, there is a huge difference between grass-fed meat and grain-fed. When grain fields are tilled, the soil releases carbon dioxide, methane, and nitrous oxide. Prairies, on the other hand, do an excellent job of sequestering carbon as they build the soil. Prior to white settlement, prairie soil contained about 9% to 12% of carbon-rich organic material. Today, the organic content of these tilled soils is about 4% to 5%. Some believe that the vast amounts of carbon released by tilling are a primary cause of climate change.

It is important to understand that much of the meat and poultry sold as organic was raised in a crowded, miserable, industrial manner. The USDA standards for organic meat allow methods of mass production that would horrify grandpa. Organic milk can come from cows that don't know what grass is. Organic eggs can come from a shed where 20,000 birds are confined.

When done very carefully, grass-fed grazing can be sustainable. It is not, however, foolproof or 100% reliable. The herds can be thinned by harsh climatic events — extreme heat, cold, precipitation, or dryness. Predators enjoy thinning the herd. Diseases can also be devastating.

Rinderpest is a highly-contagious and highly-lethal virus that attacks ruminants, like cattle. Infected animals die within six to twelve days. In the 13th century Genghis Khan brought rinderpest to Europe and China via his herds of steppe cattle. Three major pandemics hit Europe in the 18th century — 200 million cattle died. In 1887, Italians brought infected cattle from India to Ethiopia, to feed their troops. The ruminants of Africa had no resistance, and 90% to 95% of them died, including wild antelopes and buffalo. Tribes of herders lost their cattle — two-thirds of the Maasai tribe starved.

A backlash against industrial meat production is starting to gather some momentum, and a growing number of farmers are producing grass-fed beef and lamb. They are raising chickens that spend their days foraging for natural foods outdoors, and their eggs are dramatically different from factory eggs. Joel Salatin of Polyface Farms is an articulate and enthusiastic spokesman for this healthy meat, anti-CAFO movement. He has written several books.

Enterprises like the Thousand Hills Cattle Company of Cannon Falls, Minnesota are turning conventional wisdom upside down. Farmers are learning that they can convert soy and corn fields into perennial pastures and make more money producing grass-fed beef — a feat that the experts said was impossible.

Farther out on the frontier of ideas is Richard Manning. In his book *Rewilding the West*, he proposes the restoration of buffalo to their native habitat, by creating a fence-free open range, where wild and free elk, deer, antelope, prairie dogs, and all of their predators are more than welcome — and cattle are not. This sounds far more appealing than a maximum production monoculture of domesticated cattle, where the wild animals of the prairie are perceived to be a threat to profits that must be shot, trapped, or poisoned. Visionaries are fascinating people.

The lands where corn and soy are raised for animal feed today could be returned to natural grazing land. We could skip the tractors, plows, fertilizers, groundwater mining, dead zones, blue babies, and CAFOs. Our battered and abused farmlands could be returned to prairie-like grasslands where all wild beings are welcome. Projects like this are certainly prominent on Mother Nature's to-do list.

Miscellaneous Edible Critters

Until we can return to a fully sustainable way of life, it would be wise to explore modes of food production that are less harmful than the industrial food system. This section briefly examines a few options for producing protein foods.

Proteins are nutrients that are necessary for survival. Almost all of the proteins consumed in our society are produced via fish mining or soil mining. A diet based on mining is purely unsustainable. In a sustainable future, corn, rice, wheat, beans, peanuts, soy, and potatoes will have a far smaller role, if any, because of the soil damage caused by raising them. Fish mining will eliminate itself, if we don't stop it sooner — and so will soil mining.

Let's take a stroll through an imaginary mainstream grocery store, and observe the protein foods based on soil mining. The bakery department offers products made with various grains. So does the breakfast cereal aisle. There are a wide variety of pasta, bean, and peanut products. Grain and soy are substantial inputs for producing farmed catfish and tilapia, and for beef, pork, poultry, eggs, cheese, milk, yogurt, ice cream, tofu, and soymilk.

What about the protein foods not based on soil mines? There are wild-caught fish, which are typically produced via saltwater fish mining. There are wild lobster, clams, scallops, and oysters. There are hazelnuts, walnuts, almonds, and wild rice. The walnuts, almonds, and wild rice are typically grown on the deserts of central California, via water mining. There are many other protein foods that are not found in our grocery stores, and many of them are low-impact and low-cost.

There will be radical changes in the way we eat as the era of cheap energy concludes, dropping the curtains on industrial food production. Eventually, the source of our food will be the place where we live. Eventually, our diet will become healthier and more diverse. The obesity epidemic will become a comical chapter in grandpa's fireside stories about the wacky days of bad craziness.

We've already looked at grass-fed grazing, as an intelligent alternative to the high-impact process of raising livestock on grain-based feeds. We've seen brilliant Asians converting wastes into fish, avoiding both fish mining and grain-based feeds. There are many, many other options that lie beyond the sphere of industrial food production.

The path to a sustainable future demands that we devote far more brain power to thinking outside the box. The future belongs to those who are imaginative and open minded. People who are rigid about the types of foods they eat will suffer or starve. Those who can overcome cultural inhibitions will eat far better.

Many Native American tribes in the western regions ate substantial quantities of insect foods — grubs, caterpillars, grasshoppers, locusts, larvae. Insects were not merely famine food, they were a component of the standard diet, and they were considered to be absolutely delicious.

For example, fly larvae were a staple food for the Paiute people who lived at Mono Lake, California. Every summer, tons of the larvae would wash up in piles for miles along the beach. This nutritious fat meat was gathered, dried, and stored for winter food. Tribes would sometimes go to war with one another over access to areas that produced an abundance of insect foods.

Today, the vast majority in the developing world are already eating insects. In Africa, caterpillars are gathered and eaten by the ton. They are higher in protein than beef or fish. Ants are a popular food in Mexico and Columbia. The *U.S. Air Force Survival Handbook* informs us that popular meats in China include locusts, dragonflies, bumblebees, cockroaches, grasshoppers, golden June beetles, crickets, wasp larvae, and silkworm larvae.

Mainstream consumers shudder at the notion of eating insect foods, which remain quite popular in every continent except for Europe and North America. They are delicious, nutritious, organic, free-range, wild foods. A maggot salad is better food, in many ways, than a salad made with industrial chicken.

There is also insect ranching. In Thailand, there are 15,000 farmers who make good money raising locusts. Locust ranches do not require grazing land, or encourage deforestation. They have no hazardous sewage lagoons, nor are they generators of greenhouse gases. Unfortunately, these poor animals are enslaved and deprived of a life of freedom. Foraging for wild insects is more respectful.

Americans won't eat insects, dogs, cats, rats, or horses — delicious foods still enjoyed in other cultures. Outside our box, a billion Muslims and Jews shudder at the notion of eating ham or bacon. A billion Hindus shudder at the notion of eating cattle. Thinking inside the box puts limitations on our food supply, which can have fatal consequences in times of trouble.

As we've seen, in the traditional freshwater fish ponds of Asia, no grain-based feeds were used (this is far less true today). Farmers provided the fish with crop residues, kitchen wastes, manure, and human sewage. Protein-rich duckweed thrived in these funky ponds, providing food for the fish. Ducks, too, were delighted to feast on delicious duckweed (hence the name). In this manner, the amazingly clever ponds converted wastes into fish, meat, and eggs.

Ducks also love to eat slugs, snails, flies, mosquitoes, and maggots. It's quite easy to raise maggots during warm weather, all you need is a container, some stinky putrid wastes (to attract flies), and a garbage bag. Fresh cow pies are also excellent maggot incubators, as many a well-fed chicken knows. In some places, slugs and snails are annoyingly abundant, which clever people perceive to be a duck shortage.

Like ducks, turkeys eat both plant and animal foods. In proper situations, they can feed themselves by foraging, and come home at dusk to roost. Little care is needed. They can survive in a wide variety of climates. Note that the turkeys and chickens bred for industrial meat production are incapable of surviving a free-range farmyard life, they have trouble walking. Old fashioned breeds are required for non-industrial production.

Geese are grazing birds, and primarily vegetarian. They love juicy green grass, like parks, golf courses, and lawns. Geese convert things that we can't digest into meat, eggs, down, and fertilizer. They are better security guards than dogs (they cannot be calmed to silence by intruders), and they require very little care from humans.

Before they were domesticated, wild chickens were forest birds that majored in seed-eating. Chickens require more attention than turkeys, geese, or ducks. Raising them without commercial grain-based feeds is more challenging, but possible. They love maggots, and it's cheap and easy to become a successful maggot rancher.

They also love worms, and worm ranchers turn food wastes and biomass into worms and high-potency compost. The Maoris of New Zealand consider earthworms to be a great delicacy. South Africans like them fried, while the Japanese bake delicious earthworm pies. Worms are a nutritious feed for meat-eating birds and fish, and they can be dried and stored for later use.

Landless people can raise pigeons. Many pigeons are raised on the rooftops of Cairo. Pigeons are often allowed to fly away and forage for food on the surrounding countryside. Under proper conditions, foraging can be their primary source of food, and they would not have to be fed grain. Neighbors and farmers are not always delighted to see foraging pigeons, but hawks, falcons, and rats have a more favorable opinion.

Landless people also raise guinea pigs, which are a major source of food in the highlands of Peru and Bolivia. They are tame, fairly odorless, and can be raised indoors in boxes under the bed. Their primary food is grass, and they are also fed kitchen scraps. Twenty females and two males can produce enough meat to feed a family of six. The latest breeds can grow up to two kilograms (4.4 pounds).

Naturally, raising birds, bugs, rodents, or anything else on an industrial scale is expensive, polluting, cruel, and unsustainable. At the other end of the spectrum, raising critters on a small scale, low-tech, low-impact manner can work well. Some require little care, little expense, and little or no grain-based feed. They can provide us with nutritious food during the collapse. Once our population returns to a sustainable level, and the surviving wild animals recover, then we can abandon the enslavement of domesticated plants and animals forever.

In the future, humans will no longer spend their lives sitting in corporate cubicle farms. We'll have a lot more free time, and much of our work life will be involved in acquiring food, outdoors, in the beautiful real world. Let's go fishing, eh?

The National Research Council published the book *Microlivestock*, which discusses non-mainstream food animals, including poultry, rodents, rabbits, lizards, and others. It's a stimulating treasure chest of information for folks who think outside the box. As of this writing, it can be read online. The *Food Insects Newsletter* was produced by Gene R. De Foliart, and some of his papers can be found online.

Protein can also be acquired from wild plant foods, like nuts and seeds. They are the natural choice for those who are reluctant to take the life of a maggot or grub. Nuts and seeds can be acquired in a manner that is vastly less destructive than the typical mainstays of the industrial vegetarian diet (i.e., rice, beans, soy, etc.).

Acorns have long been a primary food for west coast Indians, and chestnuts used to feed many in the eastern US. Wild rice is high in protein, and can be sustainable when it grows in wild wetlands (not in irrigated California deserts). The following sections talk more about seeds and nuts.

Edible Prairies

Wes Jackson and his wife Dana created The Land Institute in Salina, Kansas. Their mission is to create sustainable agriculture. The grand vision is an edible prairie — a sustainable system of perennial plants that produces generous amounts of seeds (grain), requires no tilling, eliminates soil erosion, and actually builds new topsoil. The system would include a mix of perennial legume plants, to provide nitrogen for the other plants. The system would largely manage itself, and harvesting would be the farmer's main job.

Wes Jackson detests industrial agriculture. He thinks that contemporary till-based farming is a disease that destroys the land and the future. The primary destroyers of soil in North America are wheat, corn, and soybeans. Agriculture does not work, and it has never worked anywhere, except for a few isolated pockets. Even the venerable Amish and Mennonites have not eliminated erosion — their agriculture is also terminally out of balance. He does not see reduced-till to be a solution, because it requires periodic tilling, regular herbicide applications, and the use of heavy machinery. He thinks that it would be really cool to develop a method of growing food that caused no harm.

The Soviets also dreamed of perennial grains, and they started similar research in the 1920s. Rumors indicated that they were enjoying some success. Western scientists begged to see what they were doing, but were always denied. Western scientists worked hard to duplicate the experiments, but all efforts failed. Soviet research fizzled out in the 1960s, with no evidence of success.

Jackson believes that in 50 to 100 years there could be successful results. He wants his edible prairie of perennial plants to produce as much grain per acre as industrial agriculture produces with annual plants. This is quite a challenge. Perennials generally do not concentrate their energy on producing abundant quantities of large, nutritious seeds because they don't have to — they're perennials — they'll be here next year, whether or not they produce a single seed this year. They need to conserve energy to maintain their rhizomes and root systems, and survive long winters.

Jackson also wants the edible prairie to be machine harvestable. The seeds of the mixture of plants would have to ripen at the same time, and not promptly fall to the ground. The blend of plants used would have to be custom-tailored to the soil and climate of each location.

For many centuries, Indians in many places have nourished themselves by gathering seeds from perennial plants in edible prairies. Their system was different from Jackson's vision in three notable ways. (1) It didn't demand that perennial plants produce seeds of great size, in great quantities. (2) It didn't expect the prairie ecosystem to feed a huge and unsustainable human population. (3) It wasn't managed to enable machine harvesting.

In her important book *Tending the Wild*, M. Kat Anderson discusses in great detail how California Indians worked with edible prairies, to promote the production of wild game, nuts, seeds, vegetables, basket-making materials, and other necessities of life. They didn't simply forage for what nature provided — they worked with nature to encourage the growth of the things they needed. This work included pruning, cultivating, weeding, burning, and re-planting seeds and roots. But it required far less effort than the back-breaking labor associated with agriculture — and not a single piece of farm machinery. The soil was not molested.

The lifestyles of the Pacific coast Indians were strikingly different from those of the corn-growers to the east. Western natives lived on wild foods — acorns, salmon, deer, shellfish, berries, roots, birds, grass seeds, and so on. They succeeded at voluntary population management. They did not live in settlements surrounded by massive wooden palisades, and warfare was not a major component of their way of life. They did not experience problems from erosion or soil depletion. They did not have powerful leaders, or construct massive temple complexes for the ruling elite. In short, they learned to live in harmony with their ecosystem, and they lived in a healthy and pleasant manner.

Malcolm Margolin's book, *The Ohlone Way*, offers an inspiring description of the people who once lived in San Francisco Bay. The world described by Margolin and Anderson presents a stark contrast to the serious instability of almost all agricultural societies. They compel us to question the value of agriculture, and the need for it. They make it clear that envisioning a sustainable future without agriculture is completely realistic and highly preferable. Foraging for wild grass seeds provides nourishing food without laying waste to the ecosystem. When done properly, it is sustainable.

Edible Forests

Agroforestry is a hot buzzword these days, for good reasons. Trees are perennials. There is no need for annual plowing and planting, and all of the associated problems. Nuts are a nutritious, calorie-rich, and storable food. Many types of fruit can also be preserved and stored. Once the trees are established, the labor required for maintaining them is far less than producing annual field crops. Some trees, like olives, can remain productive for hundreds of years. Some trees can produce more food per acre than grain farming.

History informs us that agroforestry can be done in a perfectly sustainable manner. But history is equally clear that agroforestry is far from foolproof. One rule that we can carve into stone with confidence is: *Thou shalt not commit monoculture.* This is the legendary mistake of putting all of your eggs into one basket. It takes years for a tree to come into production, and it comes with no guarantees for problem-free reliability. Trees are vulnerable to climate shifts, pests, diseases, fires, vandals, and so on. Intelligent agroforestry must be diverse — spread the risk across a variety of different food-producing trees.

In the early 1990s, Brazil's coffee crop got hammered by unusually cool weather. Farmers who depended solely on coffee beans for their economic survival were wiped out by the frost. Farmers who raised a diversity of foods did not get blindsided.

Cocoa growers are suffering from a virus. The cocoa swollen shoot virus (CSSV) has been killing cocoa trees in West Africa since 1930. West Africa is home to 80% of the world's cocoa production. The virus is spread by mealy bugs. The disease slowly but surely kills the tree; it usually takes two years. In Ghana and Togo, more than 200 million infected trees have been cut down since 1948. Cocoa is Ghana's main export. The treatment is to uproot the tree and burn it, roots and all. This virus remains an ongoing problem.

Monoculture provides heaven-like conditions for pests and diseases. A cocoa tree living in a 1,000 acre cocoa plantation is far more vulnerable than a cocoa tree living in a natural forest, surrounded by trees of many species.

Another rule to carve in stone is: *Expect change.* Everything is always in a never-ending process of change. Stability is never permanent, it is simply a phase in the cycles of change. Even when an agroforest has no design defects, it is always a full participant in endless change — one year a bumper crop, the next year nothing.

On my Keweenaw ranch, I had 60 apple trees. During good years, these trees produced a lot of food, which I canned and stored for later use. But, there were three years in a row when these 60 trees barely produced at all. In 1999 it rained heavily from early May to mid June (blossom time), and there were no apples. In 2000 a heavy frost hit on May 26, when the blossoms were out — no apples. In 2001 millions of tent caterpillars came to visit, and by mid June they had defoliated every apple tree — the third autumn with no apples. Here are some of my jottings from that time:

> The tent caterpillar infestation this year was even worse than last year. It was simply unbelievable! I'd go out twice a day and fill buckets with caterpillars that I stripped from my cherry, apple, and pear trees, and the bush plums and apricots. By mid-June, there wasn't a leaf on any of my fruit trees (all of them eventually grew new leaves). They ate everything except the pines and cedars — the forests were transparent. Eerie!
>
> The presence of tent caterpillars is announced by a sound similar to the dripping of rain from trees following a shower. This dripping noise is made by leaf clippings, turds, and falling worms. When you walk, you can see the leaf clippings on the road, and the crawling worms. The caterpillars hang by threads from the trees, and they collect on your clothes as you walk, by the dozen. In Mesnard, you can't throw a dime without it coming to rest on a worm. When you walk or drive, it

What Is Sustainable

sounds like you're going over that plastic blister pack stuff — snap, crackle, pop — as the caterpillars explode.

A friend said that in past years, the tent caterpillars were so bad in Minnesota that they would sometimes close roads — they became too slippery for safe driving! The paper ran a cover story on tent caterpillars. Hundreds of thousands of acres were defoliated in the UP this year. In places, there were two to four million caterpillars per acre. They are native critters who wax and wane on a 15 year cycle. They devour green leaves for the month of May, and then make a cocoon. They rarely kill the host plant, but they make the forest look like hell.

In this scenario, not even a diverse and perfectly designed agroforest would have produced food. This emphasizes the importance of having a population that is significantly less than the theoretical maximum carrying capacity, in order to have a healthy safety cushion in the food supply. You can have total faith that the food supply will vary from year to year. Nature will happily utilize any surplus foods. Nothing is waste in nature.

Another rule that we could carve into stone would be: *Thou shalt not manage tree crops to maximize productivity.* The apples trees on my Keweenaw ranch were scattered throughout the woods, wherever the deer, bear, and coons had planted seeds in their droppings. My only management involved trimming back the surrounding vegetation, in order to provide more sunlight to the apples.

Modern industrial tree farming is an entirely different process. I was surprised when I read a European Union report detailing the ecological harm caused by growing olive trees.

Industrial olive growing is centered in Italy, Spain, Portugal, and Greece. Trees are typically grown in high density on irrigated land, which is tilled to kill vegetation. Pesticide use is intensive. Irrigation water is often produced by groundwater mining.

The report stated that when olive groves were managed for high productivity, this caused serious soil erosion problems, leading to wide scale desertification in places. There were also big problems with runoff of farm chemicals and soils into local streams.

On the other hand, the report said, olive trees that were not managed for high productivity were generally sustainable. These trees were not planted in high-density monoculture groves, the vegetation growing beneath them was not kept clipped short, and no chemicals were used on them. This method for growing perennials was about as close as we come to achieving sustainable agriculture. Not coincidentally, this method of low-density planting closely mimics the way Mother Nature does it.

In his book *1491*, Charles Mann speculates that much of the "vast unspoiled wilderness" that the first European explorers observed in the New World was actually a manmade ecosystem — carefully managed agroforests mixed with carefully managed pastures. One out of every four trees between southern Canada and Georgia was a chestnut. A variety of fruit trees were also common.

The abundance and vitality of wildlife was promoted by maintaining high quality grazing lands. This was done by regularly burning off the grasses, shrubs, and seedling trees in certain areas. There was far more meadowland 500 years ago, and the forests were quite open and park-like. When the whites arrived, buffalo roamed on open grasslands from New York to Georgia.

Mann also speculates that the Amazonian lowlands were once an abundant agroforest of nut and fruit trees, supporting a high-density population. Things took a sharp turn for the worse when Europeans arrived, and provided the natives with metal axes. The new technology spurred slash and burn agriculture, which had disastrous results.

Joseph Russell Smith published *Tree Crops* in 1929. He was a fundamentalist tree-loving eco-missionary who railed with great vigor against the hideous sins of erosion. Americans were destroying their soils faster than any nation in all history (one-third of US topsoil had been lost by the 1970s). Why not plant tree crops on hillsides — instead of plowing them, growing grain on them, and destroying the soil? With tree crops on the slopes, and pastures in the bottoms, we could enjoy sustainable agriculture. Let's do it! But American agriculture ignored his visionary bandwagon. Nevertheless, Smith's book remains a treasure chest of information on an amazing variety of crop trees.

Currently, much of our grain is fed to livestock and poultry, and growing corn and soybeans results in soil mining, water mining, dead zones from fertilizer runoff, and other serious problems. Smith recommended feeding tree foods to animals — chestnuts, acorns, persimmons, mulberries, pawpaws, walnuts, filberts, pecans, hickory, and beans (honey locust or carob). In an elegantly simple process, food falls to the ground, and the animals serve themselves. Processed grain-based feed does not need to be trucked in from distant lands, and then purchased, and then served to the critters. Tree foods could also provide vegetarians with soil-friendly alternatives to soil-wrecking soy and grain foods.

The American chestnut was once very common in eastern North America, from southern Canada down through Mississippi. These trees were about 20% of the Appalachian forest. Chestnuts were sometimes referred to as "the redwoods of the east," because they grew so large — one tree was 17 feet in diameter. They grew fast, grew straight, and produced abundant nuts every year, without fail. A big tree could drop 6,000 nuts (ten bushels or more). Chestnut lumber was prized because it was rot-resistant. The acidic bark was used to make tannin for the leather industry.

For the Native Americans, and the hillbillies who followed them, chestnut trees were the most valuable component of their ecosystem.

They fed an abundance of wildlife — raccoons, squirrels, turkeys, bears — and people, too. Hogs were turned loose in the woods in the fall to eat the nuts, and this diet resulted in exceptionally tasty pork. The nuts were also taken to market and sold to provide cash ("shoe money"). For a simple subsistence way of life, the chestnuts were a valuable gift from the gods — a generous supply of excellent food that required little work and cost them nothing. Farmers had money in their pockets and meat on their tables. It was an uncommon and very lucky situation.

There were also chestnuts in Asia and Europe. They were less valuable as lumber trees, and their nuts were larger and less tasty (often bland or sour). Japanese trees were first imported into the US in 1876. By 1900, mail order nurseries were selling Japanese and European chestnuts to customers across the country. The Asian trees carried the chestnut blight fungus, but they were fairly resistant to it — it usually didn't kill them.

The American chestnut had almost no resistance to the fungus, and the blight destroyed three to four billion chestnut trees — almost all of them — over a 40 to 50 year period, starting in 1904. Once it got rolling, the blight spread at an exponential rate. It moved southward at a rate of 50 miles per year. The spores of the fungus were spread by birds, animals, insects, woodsmen, and the wind. The blight killed the trunk and branches, but not the root systems. Millions of chestnut roots remain alive to this day. They continue to send up shoots, and the fungus continues to kill them. The blight fungus is a pathogen that can remain dormant for a long time.

The experts desperately tried to fight the blight. Sprays didn't stop it. Cutting off infected tissue didn't stop it. Cutting down every infected tree didn't stop it. The blight stopped spreading when it could find no more healthy chestnuts to infect. It missed a few trees on isolated hilltops, and in remote regions like Minnesota and Wisconsin, that were beyond the chestnut's traditional range.

The rapid die-off of chestnuts blindsided the Appalachian ecosystem — the turkeys, the bears, the deer, and the hillbillies. Country folks were horrified to watch these extremely beneficial trees suddenly die by the thousands — all of them. It felt like the end of the world, an apocalypse. The old way of life was gone forever, and the new way of life was much leaner and more difficult. Many of the younger hillbillies moved away, and many former cornfields returned to forest. By and by, Appalachia became synonymous with extreme poverty.

This chestnut story intrigued me, because it provided a powerful metaphor for the spread of civilization — a rapidly spreading pathology that destroys everything in its path, and nothing, nothing, nothing can stop it — not science, not money, not religion. It thrives until there is nothing left to destroy — and then it can remain dormant and wickedly dangerous for a long, long time.

One more rule to carve into stone is: *Stay where you belong.* The chestnut disaster is heartbreaking, because it was caused by a tree importer trying to make a buck. The list of disasters caused by innocently imported pests and pathogens is huge — bubonic plague, cholera, malaria, influenza, measles, smallpox, rinderpest, potato blight, and on and on. Countless millions have died from this simple mistake. Today, this simple mistake is being made at record-setting levels. In a sustainable world, there will be no trade or travel beyond the home region. The pathogens of the Congo will remain in the Congo, hopefully.

One last rule: *Adapt to the ecosystem around you.* Don't manage it. Don't control it. Don't turn it into a cornfield or an olive factory or a CAFO or an industrial center. Find a way to live that doesn't rock the boat.

Tikopia — An Island of Sustainability

Tikopia is an isolated island in the Pacific that was formed by an old volcano. It is one of the Solomon Islands, and it is 1.8 square miles in size (4.6 km^2). Tikopia has been inhabited for 3,000 years, and it currently has about 1,200 residents, living in 20 villages. This culture is unique because of two notable achievements. It appears that the Tikopians have developed a mode of agriculture that is essentially sustainable. They have also evolved a method of voluntary population control that has worked effectively for centuries (prior to the arrival of the missionaries).

Much of their food comes from an agroforestry system that produces fruit and nuts year round. They grow banana, apple, almond, papaya, betel nut, coconut, breadfruit, and chestnut trees. On the ground, they cultivate manioc, taro, and yams. The fertility of the soil has been maintained by 100% nutrient recycling. They also eat fish, shellfish, turtles, sea birds, and domesticated chickens and ducks. Fermentation is used to preserve starchy foods, like breadfruit, for later use.

For the first 2,000 years, the Tikopian settlement was not sustainable — their experiments with slash and burn agriculture did not work well over the long run. By trial and error and cleverness they eventually developed a system that has worked nicely for the last 1,000 years. Agroforestry is not labor intensive, so people do not have to suffer from a lifetime of endless backbreaking labor. As an added bonus, the remote location of Tikopia isolates it from crop-destroying insects and plant diseases. There are no large mammals on the island, competing for the bananas and coconuts. They used to raise pigs, but they proved to be more trouble than they were worth, so they ate them.

The food system is for subsistence purposes only — there is no commercial fishing, and no cash crops are grown. Isolation eliminates the possibility of trade, which eliminates the possibility of wealth, inequality, and the associated social frictions. They are not confronted with the temptation to acquire guns, whiskey, and televisions by logging off the mountains or over-fishing. They have no incentive to live foolishly, at the expense of their children and ecosystem. It is taboo to catch too many fish, or have too many children.

A Spanish boat arrived in Tikopia in 1606. The sailors were heartbroken to find no gold, diamonds, or anything else of immense value. The island was so small, remote, and "worthless" that it wasn't interesting to potential conquerors. What a blessing! European visitors have been generous about sharing their contagious diseases with the Tikopians, and they have thoughtfully brought them missionaries, to teach them the correct way to live and think.

Living on a tiny remote island really alters the way you think and behave, because the island is the world. They do not have the option of driving 50 miles to Safeway, if they run out of bananas. They do not have the option of clearing additional cropland, if they want to expand their food production, because there is no unused land — destroying their soil means destroying their future. They do not have the option of packing up the family and moving to the city for a factory job — their closest neighbors live on the island of Anuta, 137 miles away — a long and dangerous canoe trip across open ocean. They have to live with their mistakes. There is nowhere to run. If they live intelligently, life is good.

The Tikopian food system can feed about 1,000 people, and everyone on the island thoroughly understands this. When population exceeds the food supply, there is trouble — war, famine, social breakdown, eco-destruction, and so on. But there is no trouble when everyone has enough to eat, and feeding the gang does not damage the ecosystem. Today, billions of people have forgotten this simple, but extremely important, truth.

War is not a popular option on Tikopia, because the island is so small that everyone knows everyone. Killing, raping, and pillaging friends and relatives takes much of the pleasure out of fighting. Starvation is not popular, either. Instead, the Tikopians chose a different path. Population growth was prevented by celibacy, abortion, contraception, invalidicide, infanticide, and forced emigration by canoe (which generally resulted in drowning). Prior to the arrival of the missionaries, infanticide was common, and Tikopian culture perceived it to be perfectly normal and natural. It was absolutely preferable to the horrors of overpopulation.

Every year, chiefs hold no-growth rituals. Family planning is a core component of the culture. Depending on current conditions, families might have two, three, or four children. It is taboo to have more children than the current limit. When a child reaches adulthood, it is taboo for the parents to have more children.

Modern consumers tend to be extraordinarily self-absorbed individualists. Our way of life has no long-term future. It is almost inconceivable for us to imagine living like the Tikopians, where the wellbeing of the community is given priority, and a system of voluntary cooperation works very well. But their land produced good food, the sea was full of fish, and the people took care of one another. It sounds like utopia. It's not impossible.

Is Agriculture Necessary?

Now that we've let the agriculture genie out of the bottle, can we ever put it back again? As the population shrinks over the coming centuries, there will be less need for agriculture, and opportunities for switching to a sustainable way of living will increase. Most of human history was agriculture-free, and relatively sustainable. Many respected scientists and scholars do not hold agriculture in high regard.

In *The Evolution of Technology*, George Basalla wrote that agriculture is unnecessary. Plants are quite capable of taking care of themselves and thriving without human help.

In *Earth*, Paul and Anne Ehrlich wrote: "In retrospect, the agricultural revolution may prove to be the greatest mistake that ever occurred in the biosphere — a mistake not just for Homo sapiens, but for the integrity of all ecosystems."

In *Collapse*, Jared Diamond wrote: "Recent discoveries suggest that the adoption of agriculture, supposedly our most decisive step toward a better life, was in many ways a catastrophe from which we have never recovered."

In *A Short History of Progress*, Ronald Wright wrote: "The invention of agriculture is itself a runaway train, leading to vastly expanded populations but seldom solving the food problem…"

In *New Roots for Agriculture*, Wes Jackson wrote: "The plowshare may well have destroyed more options for future generations than the sword."

In *Rogue Primate*, John Livingston wrote: "Both agriculture and pastoralism have been dire news for Nature. Both have been the scourge of ground cover, topsoil, and watersheds; both have simplified, homogenized, and monoculturalized vast areas of the plant; both have displaced and destroyed whole populations, communities, and biotas…"

In *Rewilding the West*, Richard Manning wrote: "Sustainable agriculture is an oxymoron. It mostly relies on an unnatural system of annual grasses grown in a monoculture, a system that nature does not sustain or even recognize as a natural system. We sustain it with plows, petrochemicals, fences, and subsidies, because there is no other way to sustain it."

In *Dirt*, David Montgomery wrote: "Continued for generations, till-based agriculture will strip soil right off the land as it did in ancient Europe and the Middle East. With current agricultural technology though, we can do it a lot faster."

In *Neanderthals, Bandits, & Farmers*, Colin Tudge wrote: "The real problem, then, is not to explain why some people were slow to adopt agriculture but why anybody took it up at all, when it is so obviously beastly."

In *Tree Crops*, Joseph Russell Smith wrote: "Corn, the killer of continents, is one of the worst enemies of the human future."

In *Prehistory of the Americas*, Stuart J. Fiedel wrote: "Did agriculture bring about any improvements in the quality of life? Probably not."

In *Empires of Food*, Evan Fraser and Andrew Rimas wrote: "The invention of farming and urban civilization didn't improve the lives of most human beings — actually, it shortened lifespans, inflicted chronic malnutrition, caused disease to fester, and condemned nearly the entire population to Adam's infamous curse."

In *World Agriculture and the Environment*, Jason Clay wrote: "Farming is the single largest threat to biodiversity and ecosystem functions of any single human activity on the planet." "Proceeding along its current trajectory, agricultural production will eventually expand onto and degrade most of the habitable areas of the planet. As a consequence, most biodiversity and ecosystem services will be lost. In the worst-case scenario, life as we know it will cease to exist."

Industrial civilization and industrial agriculture have combined to elevate our society to the very zenith of unsustainability (so far). In all of human history, no era has ever come close to our abilities at ecological destruction. At the opposite end of the spectrum are the nomadic foragers, who enjoyed a relatively sustainable way of life. This was the norm for almost all of human history.

The unhealthy mode is based on the control, domination, and exploitation of the natural world. The healthy mode is about living with nature, fitting into the ecosystem, and contributing to its wellbeing. Unhealthy is big, healthy is small. Unhealthy is hierarchical, healthy is egalitarian. Unhealthy almost always includes agriculture, and healthy rarely does. Unhealthy is maximum production by any means necessary, healthy is wild, free, and happy.

Here's the bottom line — horticulture, agriculture, aquaculture, and animal enslavement are completely unnecessary if the number of mouths is kept low. If we achieve this, then it's possible to feed ourselves by hunting, fishing, and foraging — a way of life that is much gentler on the ecosystem, much less work, and much healthier. Farming is spiritually dangerous, because it involves manipulating, confining, and dominating other humans and other species — an unwholesome relationship. Masters, owners, bosses, and rulers do not bring beauty into the world.

A far healthier model for living was provided by the indigenous inhabitants of San Francisco Bay — the Ohlone. They lived in an ecosystem rich with acorns, shellfish, berries, and wildlife. They had no domesticated animals or plants. They had a complex village-based society, and their diet was purely wild foods. It was a nice complex society, too — essentially peaceful, egalitarian, and sustainable. What a terrific model! It could not be more different from the living hell of modern San Francisco — crowds, concrete, cars, and brown air.

It is not necessary for humankind to remain forever in a self-destructive way of life — indeed, it's impossible. By definition, what is unsustainable cannot endure; it is just a temporary illness. We have a broad spectrum of options. There are three huge obstacles on the path to a beautiful future: a dysfunctional worldview, an unsustainable lifestyle, and catastrophic overpopulation. All three will be remedied, one way or another.

THE TEMPORARY POPULATION BUBBLE

The ecosystems of the world are currently being pounded by a temporary bubble in the human population. We have forgotten much of our ancestors' wisdom regarding living in balance with the land base that we inhabit. We have temporarily disabled important natural population control safeguards. We have never been more out of balance with nature. In the long run, balance will triumph over chaos and disharmony. Mother Nature will never surrender to foolish human cleverness, or be destroyed by it. The existence of the human species is optional.

The Hawk and the Squirrel

The following is from my Keweenaw diary, written in the winter of early 1998.

> This week, I had an amazing experience. I was sitting in the kitchen working, and my attic was full of squirrels running, grunting, chattering, and shrieking. I went upstairs and yelled and banged on the ceiling. A couple of squirrels fled through the small hole to the outdoors and raced down the asphalt siding to the snow.
> Looking out the window, I saw a large bird sitting on a broken limb on the closest poplar. It was a hawk. And a few feet up the tree from the hawk was a very anxious and terrified red squirrel. The squirrel moved to the opposite side of the trunk and tried to sneak past the hawk. The hawk moved and stopped him. When the hawk returned to its limb, the squirrel raced down the tree, to the snow, and started racing towards the shelter of a brush pile.
> Before he got two feet, the hawk dove out of the tree and pinned the squirrel in its claws. And then it squeezed its claws into the squirrel's body. The squirrel quickly died. The hawk

spread its wings and flew away with its hot lunch. When it was gone, another squirrel ran to the place in the snow where the other had been caught. It looked all around, sniffed, but couldn't find the squirrel. It was gone — returned to the sacred circle of the dance of life.

This all happened in about one minute. I stood there amazed. I felt blessed by the experience of seeing this encounter. It's very rare to see a hawk so close — and to see it kill.

Predator and prey are ancient partners. The prey fed the predator, and the predator helped to keep the prey population in balance. They both need each other to thrive.

There are dozens of creatures who wander this land with a healthy appetite for the fresh hot flesh of squirrels. The squirrels have zero security. Every day might be their last. But they aren't paranoid or depressed or deranged. Instead, they exhibit great joy at being alive. Each and every day is a treasure to be celebrated as fully as possible.

A large herbivore with big horns can kill a lion in self-defense. And so can humans. But we got too good at self-defense. And we went beyond self-defense, by deliberately stalking, killing, and eating our predators.

Today, there are a few large carnivores remaining in the world. Even if we bred them and re-introduced them to every ecosystem, it is unlikely that humans would stop killing them — because we know how to, and we know that we can. This means that the control of the human population rests largely in human hands, and agricultural societies rarely do a good job with this. Predators are sacred. This is the lesson that the hawk gave me.

The Reindeer of St. Matthews Island

St. Matthews Island is in the Bering Sea, in the state of Alaska. The island is 128 square miles in size. In August 1944, the US Coast Guard released 24 female reindeer and five males on the island. The island was free of predators, and it had an abundance of high quality vegetation.

Naturally, the reindeer herd exploded in size — it could not do otherwise. By 1963 the population had grown to 6,000. In the absence of predators, the population was only limited by the supply of food, which ran out late in the winter of 1963-1964. At that time, the reindeer population fell from 6,000 to 50.

One limitation on population is food supply. If there is food for 50 reindeer, then there can be no more than 50 reindeer. Another limitation on population is predators. Without predators to keep the herd stable, population explosions and die offs are inevitable.

I have backpacked on Isle Royale in Lake Superior, a more balanced situation. On this island, hundreds of moose and a dozen or so wolves have been coexisting for 60 years. Thanks to the wolves, the moose don't breed themselves into catastrophe. Thanks to the moose, the wolves enjoy a healthy diet of wild organic meat, free of growth hormones and antibiotics. The dance of predator and prey is a time-proven success — one of the planet's most ancient processes.

Now, let's zoom out and contemplate our modern world. Humankind has ravaged the face of the planet in order to radically increase our food supply. At the same time, humans have fairly well eliminated the large predators that were formerly our partners in keeping the human herd under control. This is the ecological equivalent of armed and dangerous gangs of rabbits killing off the hawks, wolves, coyotes, and wildcats.

Sacred Man-Eating Predators

Most of us have zero fear of being eaten by a predator, because predators rarely appear in our modern way of life, and because our modern way of life has become so isolated from healthy wild ecosystems. Let's take a look at a world that we have forgotten.

- In her 1995 book *Spell of the Tiger*, Sy Montgomery wrote that tigers kill hundreds of people each year in the Sundarbans region on the border of India and Bangladesh. No effort is made to kill the tigers. Instead, they are worshipped. They protect the forest from eco-pillagers. Between 1860 and 1866, 4,218 people were eaten by tigers in the Sundarbans.

- A 22 July 1997 story from CNN described man-eating wolves in the region around Uttar Pradesh, India. In the area, wolves had killed at least 50 children in the last year. Some villagers blamed hyenas for the killings, not wolves.

- In *Inside Africa*, John Gunther reported that five hundred Africans were killed by lions near Ubena, Tanganyika in 1946 and 1947.

- The famous Champawat Tigress killed 436 people in Nepal and India before being shot in 1911.

- In *Man-Eaters*, Michael Bright wrote that in the last 400 years, an estimated million people have been killed by tigers in southern Asia. This book provides over 250 pages of stories about man-eating tigers, lions, sharks, snakes, crocodiles, and many others.

Predators provide us with beautiful gifts. By dining on the slow and the weak, they spare us from rotting away and dying of old age. They inspire us to pay intense attention to reality. When you walk in a forest where wild predators have managed to survive, you are a prey creature, walking meat. This is a rare and peculiar phenomenon for modern consumers, who come from a culture where humankind brutally rules over enslaved nature with an iron fist. Walking in predator country humbles us, and helps us to remember who we really are — slow, defenseless, and very tasty two-legged meatballs.

In the land of the consumers, when people go outdoors, it is usually to walk to an automobile — a climate controlled high-speed wheelchair, encased in metal and glass, with plush upholstery and a mind-numbing sound system. Most of us have become thoroughly accustomed to an indoor way of life, largely isolated from the living world, except for nature programs on television. We would not live long hiking in tiger country, busily typing text messages while blasting our brains with digital music players.

Back when we were wild, free, and happy, humans were both predators and prey, hunters and walking meat. Our legendary big brains evolved in a reality where we paid intense attention to the world around us. We were totally plugged into reality, in a magnificent state of high awareness. We were absolutely alive. We were delighting in the experience of being a pure and unspoiled all-natural human being. We can still have that experience, if we choose to.

Dersu the Trapper

Dersu the Trapper is a memorable story written by Vladimir K. Arseniev. This book documents a frontier adventure that took place in the Russian far east from 1902 to 1908. Arseniev was leading a Russian expedition to survey and map the Ussurisk region of Siberia. His guide was an indigenous nomadic forager named Dersu Uzala, a member of the Nanai tribe.

At the time, Chinese settlers were busy destroying the wilderness ecosystem. The Chinese consisted of two groups: miners and bandits. The miners mined the soil, the trees, the ginseng, the wildlife, the minerals, the fish, and the indigenous people. The bandits mined the miners, for anything of value. This was a remote region, there were almost no cops, and it was therefore a fashionable address for escaped convicts, fugitives, and other assorted outlaws.

The Chinese had also brought smallpox, which killed Dersu's wife, children, and most of his tribe. Dersu was 53 years old. He had spent his whole life sleeping under the open sky, except in the winter when he would build a sleeping hut. Dersu was a valuable guide because he possessed immense knowledge on how to survive in this land. He spoke to tigers with sacred respect, and never shot them. He saved the Russians' lives on numerous occasions.

During the expeditions, Dersu had been a master at reading the land — the animal tracks, bent twigs, nibbled grass, pebbles that had been turned, bits of fur and feather, specks of blood, the sounds and smells. To Dersu the hunter, the signs of the land provided vast amounts of important information, clearly and very obviously — like huge well-lit billboards. The civilized Russians were oblivious to almost all of this. Even when the signs were pointed out to them, they would often not be able to read the screaming headlines. Dersu was astounded. They were like helpless children. Their ability to perceive reality had been utterly crippled by a lifetime of living in boxes and thinking in boxes.

Arseniev invited Dersu to spend the winter in his comfortable well-heated home in the city. Living indoors quickly drove Dersu out of his mind, and he decided to leave. Why? "A man must live in the mountains!" Yes, a man must have healthy predators like the Siberian tiger to keep his senses sharp at all times. A man must live in a land with abundant fish and deer and nuts and berries. This is the life for which our evolutionary journey has fine-tuned us. This is the mode of living that enables us to live in a fully human manner. This is a mode of living that can work for a million years without thrashing the ecosystem. But we have a hard time seeing this — even when it is pointed out.

Sacred Man-Killing Microbes

Humankind's first line of defense against the terrible malady of overpopulation was our sacred man-eating predators: the lions and tigers and bears. For a few million years, our healthy partnership thrived. Then the farmers and herders appeared, and murdered our predators like there was no tomorrow. This was done to protect their enslaved livestock and poultry. It created totally unnatural human-dominated islands of plant and animal slavery, and these islands of disharmony had a strong tendency to grow and spread into healthy wild lands.

The loss of our sacred predators resulted in a reduction in our death rate, at the same time that farming was inflating our food supply and population. The result was a growing number of people, living in dense clusters, with poor sanitation practices, in close contact with other animals. This presented nothing less than a heaven-like environment for vermin and contagious diseases.

When Columbus came to America, he brought fierce contagious diseases, which mowed down the natives by the millions. But the natives had no catastrophic diseases to share with their new friends, despite the fact that the Americas were home to some of the biggest cities of the world at that time.

Dr. Michael Greger believes that contagious diseases simply did not exist 10,000 years ago — they were an unintended consequence of the domestication of animals, waterfowl, and poultry. Living in close daily contact with enslaved animals encouraged the spread of diseases from species-to-species. For example, wild ducks carried the influenza virus for millions of years, but it never made them sick. When ducks were domesticated, the virus was capable of spreading to other nearby farmyard animals, sometimes resulting in sickness and death. Sometimes highly lethal flu strains are created via species-to-species exchanges of mutated viruses. Humankind never experienced the common cold until we came into close contact with horse sneezes.

Clive Ponting writes that humans share 65 diseases with dogs, 50 with cattle, 46 with sheep and goats, and 42 with pigs. Greger warns us that we are encouraging the most terrible pandemic of all time — highly lethal avian influenza — by the cruel and reckless manner in which we raise animals, especially chickens. Many experts believe that this is inevitable, and could kill a billion people.

At the same time that we domesticated animals, we also domesticated plants and developed agriculture. Our population grew rapidly, we became more sedentary, and began a long learning process about sewage accumulation and the health effects of poor sanitation. Sedentary living is risky, and the larger the human herd is, the greater the risk.

Following the decline of man-eating predators, contagious diseases took over the important job of maintaining a healthy human death rate. They killed many millions of people over the centuries, which kept a leash on population growth. Let's take a quick look at just one of these diseases, the Black Death (bubonic plague).

The first bubonic plague epidemic is known as Justinian's plague. It struck in 541 AD and hammered the eastern Roman Empire. During its peak, 10,000 people died every day in Constantinople. It sporadically returned over the next 250 years. Some believe that it eventually killed between 25 and 100 million people.

In 1331, plague moved from the Gobi desert into China. Over the next 20 years, it killed 35 million Asians, including two-thirds of the Chinese population. It spread to India, Persia, and Russia.

The Black Death was delivered to Italy by Asian traders in 1347. By 1351 it had spread across Europe and killed millions. Half of the population of Europe died, and 200,000 towns and villages were emptied.

Bubonic plague was spread by infected fleas, which lived on the blood of infected rats. When the rats died of sickness the fleas moved to new hosts. In medieval Europe people lived in high density, in close company with animals, in communities that were notoriously filthy. Rats and fleas were everywhere.

The plague worked like this: an infected flea bit you, and the *Yersinia pestis* bacteria entered your blood. A pustule formed at the bite site, and then the lymph nodes in your armpits, groin, and/or neck swelled. Then bleeding occurred under the skin, creating purple blotches, or buboes. The pustule grew to the size of a walnut, and became extremely painful. Eventually your nervous system broke down, you experienced overwhelming pain, and then died — usually on the fourth day. Death was often preceded by violent fever, vomiting blood, convulsions, and bizarre body movement known as the *dance macabre*.

The Black Death was ecologically beneficial. With the sharp depopulation, nature recovered in many places. As agricultural soil mines were abandoned, forests and wildlife returned. In 1300, wolves were nearly extinct in Europe, except in the mountains of the far north and Russia. But after the plague, in 1420, wolves were seen running in the suburbs of Paris. The plague gave European ecosystems a chance to heal a bit. Unfortunately, this relief wouldn't last long — plague-resistant people survived and reproduced, civilization eventually recovered, and overpopulation returned.

The plague was also beneficial for the surviving humans. After the die-off, the air and water were cleaner. Cities weren't so crowded. Life was less stressful. There weren't enough peasants to work the land, so the lords got in bidding wars to hire them, which provided an appreciated jump in wages. Land prices fell. A number of abandoned farms were converted into pasture, livestock converted vegetation into meat and milk, and this led to better nourished peasants.

By 1666, most living Europeans had developed resistance to the Black Death. The London epidemic of 1666 was among the last of the major plague contagions. But the disease is not extinct. It's alive and mutating.

The Black Death was just one of many contagious diseases that killed many millions of malnourished and unhealthy humans living in or near agricultural regions over the centuries. For example, in 1666, one in five Londoners had active tuberculosis, a disease that would slowly kill half of them. Measles epidemics hit London in 1670 and 1674. Smallpox hammered England in 1667 and 1668. Epidemics of dysentery killed many in London from 1670 to 1672. Malaria was epidemic from 1661 to 1664. Some have speculated that malaria has killed more people than all wars and all other diseases combined.

This horrific way of life was a direct result of overpopulation, which was a direct result of agriculture and a lack of predators. If we were to turn back the clock 30 centuries, England was a far healthier place — vast forests, abundant wildlife, clean water, sacred man-eating wolf packs, and a much lighter human population.

We wiped out man-eating predators. We produced far more food. We made no serious effort to limit the human herd. Consequently, contagious diseases took over the job of population management. Humans are astonishingly clever. We are gifted at inventing things that provide short term benefits. For example, in the last 60 years, human cleverness has discovered that it's possible to obstruct the activity of life-threatening diseases by using antibiotics. Now, our population has shot through the ceiling. How clever was that?

The Brief Golden Age of Antibiotics

Antibiotics are drugs that kill bacteria, both beneficial bacteria and infection-causing bacteria, but they have no effect on viruses, like those that cause the common cold. Antibiotics are among the most important discoveries in the history of medicine.

Prior to antibiotics, bacterial diseases like tuberculosis, cholera, typhus, pneumonia, and bubonic plague could not be cured with medicines. Doctors were powerless to change the course of a patient's disease — they were little more effective than traditional tribal healers. The discovery of antibiotics was revolutionary. For the first time, doctors actually had the ability to cure life-threatening infectious diseases. This made surgery far less risky, and it enabled the possibility of organ transplants and chemotherapy. The temporary era of modern medicine was born.

The first antibiotic was penicillin, which came into use in 1943. It was discovered by Alexander Fleming. By 1945, he understood the downside of antibiotic use — it would lead to drug resistance in disease-causing bacteria. In the decades following his warning, the resistance problem has rapidly grown, and now threatens the future of these wonder drugs.

In his book *The Antibiotic Paradox*, Dr. Stuart B. Levy reports that antibiotics are used on humans, poultry, livestock, fish, shellfish, dogs, cats, aquarium fish, tobacco, vegetables, fruit trees, honeybees, ornamental plants, and other creatures. Humans are given antibiotics to combat active bacterial infections in their bodies. Animals receive about 70% of the antibiotics used in the US. Sometimes they are used to treat active diseases, but mostly they are given in low doses to encourage growth and to prevent disease.

Back in the 1940s, fish living near an antibiotic factory in New York grew to spooky proportions. This inspired research that led scientists to discover that low doses of antibiotics made animals grow faster, and up to twice as big — and cows gave more milk, and pigs produced more piglets. Therefore, it is now common (and profitable) for animal feeds to contain antibiotics.

The power of miracle chemicals is always temporary, because nature is continuously changing the game by rearranging the genes. Weeds become resistant to herbicides. Fungal plant diseases become resistant to fungicides. Insects become resistant to insecticides. Not surprisingly, infectious bacteria become resistant to antibiotics. When resistance develops, the temporary solution is to switch to a different chemical. So, the development of new herbicides, fungicides, insecticides, and antibiotics is a never-ending game — until the wizards of science inevitably run out of clever tricks, which are finite in number.

Antibiotics only kill the disease bacteria that are susceptible to them, but some survive because they have genes that make them resistant to the antibiotic. The survivors pass their resistant genes along to their offspring, in reproduction cycles that can happen every 20 minutes, creating 16 million offspring per day. Resistant genes can also be transferred to nearby bacteria of completely different species, in a process known as *horizontal gene transfer*. This means that if a Salmonella bacterium is resistant to three types of antibiotics, the DNA that provides this resistance can be passed to nearby E. coli bacteria, and to all of their offspring. Bacteria are magnificent survivors.

The more antibiotics are used, the faster bacteria become resistant to them. Typically, about a year after a new antibiotic is released, the first evidence of resistance appears. A growing number of bacteria are resistant to multiple types of antibiotics, and a few bacteria are resistant to all antibiotics — the infections they cause cannot be cured with medication, so the final line of defense is good luck and big magic. Our society is not at all cautious and conservative about using antibiotics, so the problem of resistance is growing and worrisome.

While resistance is growing, the development of new antibiotics has dramatically slowed in recent decades. Pharmaceutical corporations have little interest in these drugs, because their development costs are high, and their payback is modest (patients consume antibiotics for a week or so, not many years). Antibiotics are unique in that they actually cure disease. It's much more profitable to make drugs that merely treat the symptoms of chronic diseases, and need to be taken daily until the end of life.

At the same time, antibiotic researchers are beginning to run out of tricks for making new super-drugs — the low-hanging fruit has already been picked, and the path ahead is far more challenging. We'll have to discover new, radically different bug-killing paradigms, according to Dr. Alfonso J. Alanis at Lily Research Laboratories. He believes that we are moving into the post-antibiotic era. Our ability to keep making powerful new antibiotics in the coming decades is in serious doubt.

When used full-strength, antibiotics are more likely to wipe out all of the infectious bacteria. When antibiotics are used at less than full strength (i.e., most animal use), more bacteria survive, which increases the odds that they will develop resistance. Many of us are receiving low doses of antibiotics every day, in our meat, milk, water, fruits, and vegetables.

Animals excrete about 90% of the antibiotics they ingest, and much of their excrement is used for fertilizer on cropland. Some antibiotics do not promptly biodegrade, so they may accumulate in groundwater.

The US Department of Agriculture has found that plants absorb antibiotics from the soil, where manure has been applied. Traces have been found in fruits, vegetables, and vegetable leaves. Where the soil contained higher levels of antibiotics, so did the plants. Further research is underway.

Composting manure takes time and labor that commercial farmers generally do not have. When manure is composted, the heat developed during the process may degrade some of the antibiotics in the manure, but not all.

In fact, some antibiotics will survive rigorous cooking. Researchers in Holland tried to degrade the antibiotics in meat by using an intense 4½ hour, three-step process that utilized temperatures up to 134°C (273°F). Of the 16 antibiotics tested, seven remained active to varying degrees (15% to 80%).

Today, there are almost no living doctors who worked in the pre-antibiotic era. Modern doctors take antibiotics for granted. They are used to having miraculous disease-curing power that was unimaginable 70 years ago. The growing number of drug-resistant infections totally spooks them. It drives them crazy to helplessly watch a patient die from an ordinary-looking infection, resulting from a skinned knee or a paper cut. They are losing their control, and the future of antibiotics is dim. Research is waning, and resistant pathogens are growing.

In *Rising Plague*, Dr. Brad Spellberg wrote that we are never going to win a war against microbes. The microbes will inevitably develop resistance to any and all new pharmaceutical weapons. It's only a matter of time. No amount of money or magic can fix this. How much longer will our antibiotic defense system remain effective against bacterial diseases? Without a well-stocked arsenal of powerful antibiotics, modern medicine loses much of its magic. Many more patients will die from infections, as a result of ordinary injuries, medical treatments, and exposure to sick people. This raises many important questions about the future of surgery.

In addition to bacterial diseases, there are also diseases caused by viruses, like AIDS, Ebola, smallpox, dengue fever, polio, yellow fever, and influenza. Influenza is especially worrisome to health experts, because it readily mutates, defying our persistent attempts to invent a silver bullet vaccine. Devastating pandemics, like the 1918 Spanish flu, remain a very real possibility.

Will we return to an age of contagious diseases? Dense populations beg for illness, madness, and misery, in many ways. If we were to engage in some clear thinking, and imagine more sustainable ways of living, we could move with greater speed toward a far more enjoyable future for those who come after us.

The most effective way to prevent contagious disease is to deliberately and aggressively eliminate overpopulation, return to a nomadic way of life, avoid trade and contact with outsiders, and stop enslaving and living in close contact with other animals. In other words, return to the traditional human way of life, and live like our ancestors lived.

Reverend Malthus

Thomas Robert Malthus (1766-1834) was an English pastor and scholar who lived in an era of explosive change. Population was skyrocketing, and the Industrial Revolution was building factories like crazy. Cities were jammed with growing hordes of overworked, sick, hungry, dirty, poor people. It was a time of chaos and suffering, and this was a matter of concern for Reverend Malthus.

Malthus recognized that population growth led to increased misery, since it tended to grow faster than the food supply. He wrote a book to explain this, and became a famous name in history — and the source of a fierce and enduring controversy. Even today, people are lined up to kick Malthus.

In a nutshell, Malthus believed that, *without checks*, population growth would outstrip food production. For example, a farm village is able to double its population much more easily than it could double its food production. He believed (correctly) that it was possible to increase food production, but he also believed (correctly) that food production could not be increased, without limit, forever. Thus, in theory, there are limits to the maximum population possible worldwide — a perfectly reasonable idea.

But population growth is never "without checks." It is impossible for 100 people to survive on a food supply adequate for 50, so people will starve. Starvation is a check. Malthus cited three types of population checks:

- A *positive check* results in an increased death rate (i.e., hunger, disease, war).

- A *preventative check* is a lower birth rate (i.e., abortion, birth control, late marriage).

- *Moral restraint* has to do with delaying marriage until you can support a family, and remaining celibate until then.

Malthus had at least three groups of critics. One group was intelligent people having ordinary common sense. They found Malthus to be annoying for tediously belaboring the bloody obvious. Why write a 500 page book to explain an elementary truth?

Another group took issue with the notion of moral restraint. Malthus believed that this was the only effective way of reducing poverty. He opposed poor laws (a welfare system) because providing relief for the poor was not the cure for poverty, it provided life support for perpetual poverty. Critics see this as mean-headed eugenics, aimed at eliminating the poor. This is a prickly issue. How do you make poverty disappear? Poverty has been a component of all civilizations.

The loudest group of critics consisted of utopians, who were convinced that industrial society was on the fast path to paradise on Earth. They included Karl Marx, Friedrich Engels, William Godwin, and Marquis de Condorcet — fiendish optimists who suffered from a painful blind faith in science and progress. They believed that human achievement had no limits whatsoever. Thus, population growth was not a problem. In view of the immense squalor and misery of his era, Malthus thought that the utopians were utterly daffy. His terrible heresy was to mock the cherished myth of progress, to express views based on common sense and clear thinking.

There are a large number of critics who (incorrectly) think that Malthus was a raging doom pervert, prophesying an immanent catastrophe in the near future (which he did not). Since we're still here 200 years later, and since there are seven times as many of us, then Malthus was wrong, they say.

To the utopians, Malthus was a heretic for suggesting the existence of limits. But the mass atrocities of the 20th century effectively hosed down the vivid fantasies of utopian progress. Yet, the myth persists, and continues to obstruct discussions about healthy and intelligent change. Our society remains committed to perpetual growth at any cost. This is irrational and self-destructive. It is not clear thinking.

Malthus smiles, pained, but also amused by the absurdity. As food production peaks and declines, checks will do what needs to be done to thin the herd. Overpopulation is a self-solving problem. If humankind fails to limit its numbers, nature will step in and do what needs to be done. It's really that simple. So, who will end up doing the uncomfortable business — a clear-thinking and compassionate humankind, vicious genocidal zealots, or the strong, efficient, and merciless Mother Nature?

The Population Taboo

Today, we're at seven-point-something billion and still growing rapidly. The challenges are growing:

- Our ancient predators are few in number and getting closer to extinction.

- Our food production system is close to maxed out, and is starting to crumble. Productivity gains are slowing, while ecological decay is growing. There are no silver bullets in the pipeline.

- Our medical system is temporarily keeping the death rate far lower than what would be normal during a catastrophic population explosion.

- Our energy supply is no longer cheap, and costs are certain to rise. In many ways this will hamper our ability to maintain a bloated population.

- Our population continues growing by millions per year. The birth rate is slowing a bit, but not fast enough to avoid exceeding the food supply in the near future.

Reproductive *freedom* is fiercely defended, while far less attention is devoted to reproductive *responsibilities*, and our obligations to future generations. So, what is the big question that our policy leaders have circled their wagons around? "How can we feed ten billion?" What can we do to boost food production even more? This is the Utopia Express. We can't stop growing! The Technology Fairy will certainly rescue us once again. Nothing is impossible for science. Malthus was wrong.

Is this the path to healing? Is the primary challenge confronting humankind actually a matter of food production? We don't need to feed more, we need to breed less. How?

Is this a good time to bring even more humans into the game? The biological drive to love and nurture children is powerful. There are children in every community who do not enjoy abundant (or even minimal) quantities of love and nurturing. This provides a wonderful opportunity for caring childfree adults to help kids in need, without bringing additional new people into the world, and without importing orphans from faraway places.

If you were an unborn soul, would you really want to be brought into this mad world? Wouldn't you rather sit in the waiting room for unborn souls and read glamour magazines for a few hundred years? Wouldn't it be better to wait and be born into a world where humans lived in peace and harmony with all of the family of life? It's not now or never, is it?

What does Earth really need at this point in time? More salmon, more redwoods, more rhinos, more condors, more whales, more tigers — more of damned near *everything*, with one mere exception — domesticated organisms. With each passing year, I become more and more thankful that I never brought children into this culture. I would not want to be born now. I imagine that the spirits that might have become my children instead became chickadees, ravens, and blackberries, my allies in this world, my sacred friends.

In our culture it is taboo to talk about the "P" word. Population is a serious issue, but governments can't talk about it. Churches can't talk about it. Even environmentalists can't talk about. Environmentalists used to talk about it, but then their flow of contribution dollars sputtered and wheezed, so they quit. The "P" word is an intensely emotional issue, so we have made it taboo. At a time when a huge temporary population bubble is pounding the planetary ecosystem to pieces, a taboo on discussing population is not at all helpful.

Sharply increasing our food production is a problematic goal, because it's impossible. Sharply reducing the human population is another problematic goal, but it is possible, it makes far better ecological

What Is Sustainable

sense, and it is far more beneficial for the generations yet-to-be-born, of all species. Remember that population reduction is inevitable, one way or another. This bubble is temporary. The variable here is the level of discomfort during the reduction process.

What is the sustainable population? It depends. There is not an exact number that provides the correct answer to this question. Some experts claim four billion. Others say two billion. But how do you feed two billion people without soil mining and fish mining? There were two billion people in 1925, and it was anything but a sustainable utopia. The experts know this. But if they gave us realistic numbers, they would immediately be sent to a lunatic asylum, and their professional careers would be over. This brings to mind an old Yugoslav proverb: "Speak the truth. Then run."

The sustainable population should not be thought of as the *maximum* number of people who can be kept alive without the practices of soil mining and fish mining *under ideal conditions*. The sustainable population is one that thrives without mining, and maintains an adequate safety cushion of food to survive multi-year droughts, deluges, extreme temperatures, fire, insect invasions, plant diseases, etc.

But this is the wrong way to look at it. Trying to define the ideal global population is a pointless exercise in industrial thinking. The world is not a huge factory. We are not sitting at a switchboard in the control room, eyes fixed on the population meter, regulating the global births and deaths with knobs and levers. We are not the masters of the world. Sustainability is a local thing. Does everyone get enough to eat? Are your food sources local? Has your community learned how to live in balance with the living land around you? Is your community contributing to the health and vitality of the ecosystem around you? These are better questions to ask.

In modern times, there have been a number of experiments in population reduction. Hitler, Stalin, Pol Pot and others tried mass extermination. It worked, but it was coercive, cruel, and did not enjoy widespread popular support. Not recommended.

China's One-Child Policy was introduced in 1979. In certain segments of society, families are limited to having just one child, with some exceptions. It is a compulsory program, not voluntary, and it has led to more than a few ruffled feathers, but most Chinese support the program. It has not yet succeeded at stopping population growth in China, but by 2008 it had prevented 300 to 400 million births. Compared to the predictable horrors of unrestricted breeding, this system is intelligent, responsible, and commendable.

Despite its drawbacks, the Chinese system is vastly superior to the population reduction program in America, which has yet to be initiated (or even contemplated), due to the opposition of militant fundamentalists, and our culture of militant adolescent authority-hating individualism.

The fundamentalist Islamic government of Iran also does a far better job of family planning than the US. From 1987 to 1994, Iran cut its birth rate in half. With the full support of religious leaders, the government vigorously promoted free contraception products. They were the first Islamic nation to legalize male sterilization. Young couples are required to attend a birth control class prior to getting a marriage license.

Voluntary population management would be the preferred approach. In fact, it was the norm in most societies for most of human history. It's not impossible — it's intelligent. In the worldviews of many simpler cultures, there was no evil or cruelty involved in population management — it was simply a necessary component of sustainable living. Infanticide was considered to be perfectly normal, acceptable, and proper. They had a *different value system*, and — not coincidentally — a much healthier ecosystem. Everyone realized that avoiding overpopulation was the lesser of evils. It was less painful than perpetual war, mass starvation, or turning their ecosystem into a devastated wasteland.

People knew how many mouths their region could support in a stable and conservative manner, while maintaining a large safety cushion of set-aside food for lean years. They knew that too many mouths certainly led to painful problems, so they actively managed their herds. Here are some notions I learned while reading a mountain of anthropology books:

- Among nomadic foragers, it was common for the elderly and infirm to be left behind when they became a burden on the clan. If the clan stopped moving, the clan would starve.

- Infanticide was common. It was chosen if the baby was crippled or deformed, if the current food supply was tight, if the mother was still nursing a previous sibling, if twins were born (usually one was kept), or if the infant was simply unwanted. In strained situations, up to half of the newborns were not named, fed, and welcomed into the community.

- In some clans, unwanted pregnancies were terminated via the use of herbs.

- The possibility of conception was reduced by abstaining from intercourse, by couples living apart, by using herbal contraceptives, and by extended nursing (three to four years, or more — occasionally into the teen years).

- Some believe that wild women were far more in touch with their bodies, and that conception could somehow be prevented by an act of will. For example, whites who spent 25 years with the Mbuti Pygmies never once saw an unmarried girl become pregnant, despite the fact that pre-marital sex was common, accepted, and encouraged. They saw no evidence of abortion or contraceptive use. This mysterious phenomenon was reported among other nomadic forager groups, too.

- When the hunting was bad, people were sacrificed. This was usually done with children or elders, preserving the skilled, strong, and healthy adults. The old would go first, and then the young.

Infanticide was also widely practiced in modern civilizations, because it was less painful than watching children starve to death. In Europe, infanticide was a common practice until the mid-19th century. Throughout most of the Christian era, the church largely ignored the practice of infanticide, despite officially banning it (this may be because it was usually the church's responsibility to care for the poor).

With the coming of the Industrial Revolution, many rural European peasants moved to the cities, where it was more difficult to dispose of unwanted infants. To meet this need, Napoleon established government-run foundling hospitals, where parents could anonymously give up unwanted children, by depositing them in a turntable device. From 1817 to 1820, 20% to 30% of all French infants were abandoned at foundling hospitals, where 90% of them died within a year.

In England, the *baby farm* industry was created to provide infanticide services. If a woman made a generous contribution, the baby farmer would agree to "care for" her child — the mother understood that she would never see her child alive again (the kids were starved, poisoned, strangled, etc.). In the 1870s the industry was exposed, and a crackdown on baby farming began. This led to the creation of government-mandated birth certificates, which made infanticide more difficult.

In a number of regions, the right to marry was restricted. Only those who had the means to house, clothe, and feed a spouse and family were allowed to marry — essentially, only those who inherited farmland from their parents. Landless bachelors and spinsters were not permitted to marry. They often remained on their family farm, spending their lives as helpers. Others drifted off to find work elsewhere.

In recent decades, the process of family planning has gotten much easier. Effective contraceptives and legal abortions (in progressive countries) have helped reduce the number of unwanted children. Overpopulation is a major crisis in which every man and woman, in every land, gets to vote — for or against — with their life choices.

As we move toward a sustainable future, the ideal would be something similar to the system used in Tikopia (see page 213). The core entity in their society was the community, not the individual. Their system was cooperative and voluntary. It was built into their culture, making it normal and accepted. Their culture was rooted in a functional, time-proven worldview, which was based on living in harmony with the land around them. It worked, and it worked well. Cultures in which the survival of every fetus and newborn is deemed to be more important than the wellbeing of the community and the ecosystem can never be healthy and stable.

Human overpopulation is a serious problem, as is the overpopulation of things (televisions, cars, fast food restaurants). So is climate change, toxic pollution, soil destruction, deforestation, mass extinctions. You know the list. One problem underlies all of these: the civilized worldview. It is unbalanced, dysfunctional, and it actually motivates us to live in a destructive manner — and call it progress. *This worldview is the mother of our problems, and if we don't abandon it, we're doomed.* It has deep, ancient roots. It's incredibly powerful. The healing process will not be quick or easy.

HEALING OUR WORLDVIEW

I'm amused and embarrassed by the title of this chapter, *Healing Our Worldview*, which reveals my cultural upbringing. It springs from our technological mindset. We can control anything. Step-by-step instructions for updating our worldview are provided on the following pages. Any questions should be directed to the Support Desk. Right?

Realistically, cleansing our consumer worldview of dysfunctional beliefs, in a planned, orderly, and efficient manner — radically altering the thinking of billions of human minds, instilling a profound sense of reverence and respect for the Earth, eliminating materialism and self-centeredness — is an absolutely ridiculous idea.

But, as the fossil fuel era fades, we can be certain that the unfolding collapse is going to radically alter our worldview, because our mode of living is going to undergo radical changes. The worldviews of the post-consumer era are not going to be designed and implemented by highly-educated Worldview Management experts working at the nerve centers of the global civilization. The alteration process is going to be bottom-up, and happening everywhere. All changes will be born and nurtured in the hearts of ordinary individual people.

You are hereby formally invited to immediately begin making meaningful and significant contributions to the most important healing process in all of human history. Your gender, age, race, height, weight, sexual orientation, income level, education, religion, and political connections are irrelevant. But your heart and imagination are tremendously important.

We were all born on a beautiful and sacred planet. We need to remember how to live and think as though this planet, our home, is beautiful, sacred, and worthy of our absolute respect. The first step in this healing transformation is to begin dreaming the new way of life, and envisioning the journey. If we can imagine a genuinely healthy destination, and move in that direction, we'll waste less time wandering aimlessly, or desperately clinging to sinking ships. It's a highly creative process, and there are countless good approaches.

Or maybe the first step is ripping off our blinders and taking a cold, hard look at our poisonous way of life. It's difficult to give up self-destructive habits that we perceive to be desirable or necessary. Imagine looking at your life through the eyes of your wild ancestors. We need to begin freeing ourselves from social chains, and stop being slaves to fashions, trends, and compulsive shopping. We need to stop marching to the master's drumbeats. We need to move toward love, hope, and celebration — healing.

A primary objective of this book is to serve as a spiritual laxative, to loosen up constipated hearts and imaginations. Life in consumer society can get us uncomfortably plugged up and bloated. This chapter is a grab bag of ideas, sort of a do-it-yourself kit, without an instruction sheet. It contains stories and ideas that you may find stimulating and inspiring, or not. Take what's useful, if anything, and leave the rest.

A Tale of Two Mental Worlds

The birth of domestication led to a great rift in human history. On one side of this rift was Fairyland, where all of the wild ones lived together in a relatively balanced and elegant manner. On the other side was slave country — civilization — where domesticated plants and animals were controlled and exploited by humans who fancied themselves to be masters and owners.

Many pages ago I suggested that all living things, both plants and animals, have common ancestors in ancient one-celled creatures. All of today's living things are their descendents. We are all related. This is a *genetic* web that binds all life together into one large family.

Similarly, and not coincidentally, many believe that there is also a *spiritual* web that binds all life together. At the core of our being there is a spiritual power that different cultures describe with different names. Martín Prechtel, a Mayan shaman, calls this deep core the *indigenous soul*. He describes the indigenous soul as an ancient place of spiritual origination that exists inside every living thing, both plant and animal.

In other words, what humans are at their deepest core is very much the same as what the roaches, oaks, bears, whales, frogs, and vultures are at their very deepest core. All life is alike at core, but in the morning we put on different clothes, move on different paths, and do different work in the world, according to Prechtel.

Among wild folks, the indigenous soul is conscious and a constant presence in daily living. In civilized folks it is largely unconscious and unknown. Prechtel sees that modern civilized folks suffer from a painful inner struggle between the indigenous soul and the rationalist mind. This pain leads to spiritual illness, which can then lead to physical illness. He sees the healing process as one of consciously remembering and recovering the lost indigenous soul, and integrating it into our lives.

It is obvious that our modern mindset has profound shortcomings with regard to living in balance with the Earth. Many thinkers have given this problem serious thought, including some who are less mystical than Prechtel.

Carl Jung, the brilliant Swiss psychiatrist, also recognized the existence of two mental worlds in painful conflict. He called the ancient deep layer the *original mind*, or the *unconscious*. This mind still exists in us but is often hidden beneath the *conscious mind* — the newer layer of the mind, and the engine of *directed thinking* (logic and reason).

The original mind is a world of images, feelings, memories, dreams, and myths. We access this unconscious mind in our dreams, in our daydreams, on psychedelic voyages, and whenever we pause for a moment in our directed thinking.

Directed thinking is thinking in words. It is a tool for complex communication with others, and it is largely a modern acquisition. While the endless flow of imagery in the original mind is effortless, directed thinking requires concentration and is difficult and tiresome.

Directed thinking does *not* make us more intelligent than primitive people but it does make us more unstable and less content.

Primitive people were sacred beings living in a whole and sacred world. By feeling filled with divine energy they had more strength to deal with life's uncertainties. Jung said that modern people have been cut off from their ancient roots by the rise of directed thinking. Our use of reason has stripped nature of its rich psychic power, making our world cold and barren. This psychic alienation and emptiness made us self-destructive. Directed thinking enabled the development of technology, as well as our compulsion to control nature. Jung said that the lives of modern people are ruled by the goddess Reason, and that reason is a powerful and tragic illusion.

Different societies have varying levels of consciousness. Primitive societies are less isolated from the unconscious, and therefore have an easier time integrating ancient instinctual information into their lives. Primitive societies have a rich collection of myths, songs, rituals, and ceremonies that provide meaning in their lives, and directly connect them to their ancestral dreamtime. They have strong and healthy roots that link them to their place and past. This is good. Rather than observing nature as an outsider, they are blessed by being able to fully experience the natural world. They feel at home in nature. This is good.

Modern consumers, on the other hand, are largely disconnected from place, past, and the natural world. Our powerful conscious minds are largely unaware of the million year old unconscious mind within us, and the wealth of ancient wisdom that it contains. We have become rootless super-thinkers, wizards of technology. Rootlessness makes us far more susceptible to psychic epidemics, like Nazi Germany, or consumer society. We have become a society of extremely powerful loose cannons.

Our technological skills have led to explosive population growth. Jung noted an important lesson from history: bloated populations typically exhibit higher levels of stupidity, and reduced morality and intelligence. In 1916, he was shocked by the deaths of 500,000 people in a war between rational, well-educated, Christian nations. In World War II, the spectacular brutality of the Germans shattered his belief in the essential goodness of humankind. The age of science and technology was producing poison fruit.

Modern man is clearly not a finished masterpiece, far from it. Jung understood that consciousness was not purely evil, it was a two-edged sword. On the bright side, it could help humans live their lives more smoothly. On the dark side, it gives us the power to build hydrogen bombs and dams. Eliminating consciousness from humankind is impossible.

Jung's vision was that the conscious mind could be expanded and improved so that it was not disconnected from the unconscious. But poor modern man is roaring with fears and apprehensions. Our worldview is in complete conflict with Mother Nature, and this is very painful. We will do almost anything to avoid self-examination, because we are afraid of what we'll find. So, we fill our lives with distractions and fantasies in a futile effort to escape from the pain of reality. Jung was saddened by the number of rational people who remain abysmally unconscious. Many avoid using their minds as much as possible, or use their minds with impressive foolishness. Some are almost zombies, having no imagination at all.

Escape and denial require far less effort than growth and healing, but they lead us over the edge of the cliff, to oblivion. Jung believed that growth and healing were possible and worth the effort. We needed to reconnect with the natural world, and with our ancient natural mind.

Jung provided us with a list of recommendations for inner healing — *spend more time with nature, live in small communities instead of cities, work less, engage in reflection in quiet solitude, reconnect with our past, avoid distractions, pay serious attention to our dreams, and simplify our lifestyle.*

Jean Liedloff, author of *The Continuum Concept — In Search of Happiness Lost*, also saw the two mental worlds. Her life's journey included numerous visits to wild and free cultures in South America. The Tauripan people of Venezuela were the happiest people she had ever met. All of their children were relaxed, joyful, cooperative, and rarely cried — they were never bored, lonely, or argumentative. Be aware that she was not a dreaming romantic — she actually lived with these people. Other anthropologists report the same thing in other wild cultures. Prior to civilization this was apparently the human norm.

The Yequana people seemed unreal to Liedloff because of their lack of unhappiness. As her expedition was moving up a challenging jungle stream, she noticed that the Italians would get enraged at the slightest mishap, while the Yequana just laughed the struggles away. Their daily life had a party mood to it.

Liedloff said that the new *conscious mind* (intellect) can handle just one idea at a time, so it is a poor tool for processing complex information. Our ancient *instinctual mind* can process numerous threads simultaneously, and consistently generate holistic responses appropriate for a complex and dynamic reality. She thought that in happy cultures, instinct dominated and intellect was subservient — and in unhappy cultures intellect was the powerful dictator.

She believed that civilized people are lost, infantile, alienated, damaged, unhappy, and living in constant misery. We spend so much of our life energy on damage control and struggling with anxiety and suffering. Happiness is not our natural state.

She said that we live for the future, always hungry for more, convinced that "if only I had (fill in the blank), then I would be complete and happy." Unfortunately, getting (fill in the blank) never leads us to peace and contentment. Compared to the joyful wild people, Liedloff concluded that, on many levels, we might be better off dead.

Peter Freuchen spent a lot of time with the Eskimos and married into their culture. He believed them to be the happiest people on Earth. Colin Turnbull spent years with the Mbuti and was amazed by their joyful way of living. He said that Pygmies laughed until they could no longer stand, then they sat down and laughed.

The anthropologist Jack Weatherford has spent time with head hunters and cannibals, and found them to be some of the kindest and most caring people on Earth. But a great place to observe violent savage barbarians was on the streets of his hometown, Saint Paul, Minnesota.

When Western researchers visited the Trobriand Islands to study neurosis among primitive people, they couldn't find any. When other researchers tried to study depression among the Kaluli of New Guinea they found zero depressed people. Do you see a pattern here?

I met Walter Bresette on Friday, October 24, 1997. On that day, the Hong Kong stock market took a sudden dive, which inspired other stock markets in Southeast Asia to dive, which inspired the US markets to dive. The Dow Jones index plummeted 186 points, to 7847.

Walter, Vern, and I were walking through the Michigan Tech campus that evening, and Walter was delighted by the news of global financial disaster. He was eager to return to his motel room, turn on CNN, and "watch the whole <bleeping> thing collapse." He deeply resented the <bleeping> thing because of how it had savagely damaged his land, his people, and his culture in the last 150 years. Obviously, the <bleeping> thing had only one destination — oblivion — because it was 100% unsustainable. Walter hoped that this was the day he had been praying for — the death scream of the monster.

On that same evening, millions of other people had a very different take on the very same news. They were squirming with fear and anxiety. Their fantasy world of waste and excess seemed to be heading for the rocks. If this trend continued, they could lose their McMansions, resort cottages, SUVs, ATVs, RVs, snowmobiles, and their life savings. Horror!

And so you can see that it's all a matter of perspective. Our values, morals, perceptions, thinking, and behavior are profoundly influenced by the worldview of the culture that we were raised in.

What is a Worldview?

A worldview is a system of beliefs that provide the foundation for human thought processes. Our worldview determines how we perceive the world, how our society operates, and how we live. Speaking *very generally*, there are three types of worldviews:

- Nomadic foragers are wild people who live in a wild world. They are animists who have profound reverence and respect for their home ecosystem. The basic unit of wild societies is the clan. They enjoy a high-leisure lifestyle and low rates of contagious or degenerative diseases. They live directly in paradise, here and now.

- Farmers are domesticated humans who subdue the wild world to produce domesticated plants and animals. Over the last 5,000 years, they have almost eliminated the nomadic foragers and the wild world. The basic unit of farm society is the village. Farming is a highly insecure way of life in which disease and warfare are common. In their religions, a life of backbreaking work is rewarded with cosmic bliss in the afterlife.

- Consumers are people who inhabit a densely populated manmade environment where life is centered on producing and consuming goods and services — working and wasting. The basic units of consumer society are the nuclear family or independent individuals. Consumers commonly suffer from degenerative diseases and mental illness. Paradise is experienced in consumer orgasms — brief surges of ecstatic fantasies following the acquisition of consumer products or entertainment experiences.

The worldview of wild Pygmies perceives their forest to be a place of immense holiness and goodness. The worldview of consumers — the dominant worldview today — perceives that same forest as being a natural resource, something that should be converted into wealth by destroying it. It could care less about the wild Pygmies that have lived there since the dawn of time. For the Earth to heal, the consumer worldview must die — and it will, one way or another.

We inherit our parents' worldview, and we usually carry it to our graves. A missionary once boasted: "Give me a child for his first seven years, and I will have him for life." There is much truth in that. Usually, our worldview is more or less invisible in our thought processes. Worldview beliefs are assumed to be eternal truths — ideas that are never questioned. The Earth is flat, more is better, slaves are valuable, forests are natural resources, and humans are number one. Right?

Once upon a time, someone somewhere once said that environmentalists are like firefighters in a town where all of the children are taught to be arsonists. We set fires every time we reach for our credit card. Environmentalists can race from fire to fire, dousing the flames, but their activities have no effect on the overwhelming strength of the arsonist way of life. No learning or awakening ever takes place.

As a European-American, my ancestors have been domesticated for many centuries. The worldview of domestication has permeated every molecule of my culture and its knowledgebase. It is deeply rooted and incredibly strong. It is the air we breathe. A worldview is not something that can readily be removed and discarded, like a soiled nappy. It cannot be uninstalled with a mouse click. It cannot be healed during a weekend workshop with illuminated gurus or shamans. Changing a sick worldview is a healing process that will likely take generations.

There is a touching story in Jared Diamond's book *Collapse*. It concerns the Norse settlement in Greenland, which lasted 450 years. This colony starved to death, and went extinct, despite the fact that the fjords and coast were loaded with many, many fish, seals, and whales.

They had no wood with which to build proper Norse-style boats. They chose not to make boats made with animal skin, like the Inuit used. The Inuit were masters of Arctic survival, but the Norse looked down on them as being savages and heathens. Imitating savages was undignified.

The settlers destroyed themselves by clinging to their traditional Norse culture, which was completely unsustainable in Greenland's ecosystem. If you wanted to be a farmer and herder, dress in woolen garments, and live like someone in Norway, you were not going to make it in Greenland. If they had chosen to adapt the sustainable time-proven Inuit way of living, they could have survived and thrived. But this would have required a radical change in their mode of thinking and living.

At the same time, please remember that a worldview is not invincible or invulnerable. It is merely a collection of ideas invented by human minds, and humans are quite capable of questioning, learning, and growing — especially in times of social instability. Hopefully, the healing process will be accelerated by the fact that we are racing into a turbulent era, driven by numerous catastrophes that originated in the farmer and consumer worldviews. Conventional thinking is going to lose a lot of its shine, inspiring many to explore other alternatives. Questioning, reflection, and imagination are powerful tools for change.

Nazi Hope Fiends

Adolf Hitler spent the last weeks of his life in an underground bunker in Berlin. It had reinforced concrete ceilings that were eleven meters thick (36 feet). The Red Army, Brits, and Yanks were working day and night to bring a bloody end to the Third Reich. They made steady progress every week.

Meanwhile, Joseph Goebbels, the Reich Minister of Public Enlightenment and Propaganda, kept the German people from falling into despair. The new technology of radio broadcasting allowed the Nazis to make a quantum leap in mind control sciences. When Joseph

Goebbels spoke, every German listened. Never before had a ruling elite had the power to communicate directly with millions of subjects, on a daily basis. In his radio rants, he repeatedly assured the public that a German victory was certain, close at hand — and they believed him!

Albert Speer was the Reich Marshall of Industrial Production. In late March of 1945, just weeks before the fall of Berlin, Speer was driving around in farm country, and he got a flat tire. By this time, Speer clearly understood that the war was lost, and that the end was just a matter of time. Germany's cities and industries had been bombed into wastelands of twisted metal and smoking rubble. But the farmers he met that day amazed him. They still had tremendous faith in both Hitler and the war effort. They believed that Hitler was deliberately letting the enemy forces pour deep into Germany — it was all a cunning trap. Any day now Hitler would unleash a new and terrible secret weapon, the enemy would be crushed, and Hitler would claim victory. There were many people high in the Nazi government who also shared these beliefs.

This is like the people running around today who have absolute faith that the Technology Fairy is going to give us a large arsenal of miraculous secret weapons — cheap and abundant alternative energy, sustainable industry, sustainable logging, sustainable agriculture, sustainable mining, pollution-free cars, totally amazing wonder crops — any day now. The economy is going to recover, we'll enjoy unimaginable prosperity, ten billion people will have plenty to eat, and everyone will be able to shop like crazy, with unlimited credit, until the end of time. The future will be just great, and it will require absolutely no sacrifices in the way we live. And many believe this, including top government officials.

Our souls are twisted by denial — "If our way of life was so bad, then why is everyone in the world trying so hard to live like us?" We can't go back. We can't stop progress. Once the genie is out of the bottle, you can't put it back. There's nothing we can do — one person changing makes no difference. The kids will figure something out.

Technology is constantly improving. The environment is better now than 30 years ago. Environmentalists exaggerate problems to extremes. The world is OK. There is nothing to worry about. Be optimistic. Have a nice day...

Denial is the mother of apathy, and apathy blocks the path to growth and healing.

Traudl Junge was one of Hitler's secretaries, and she was in the bunker when the Führer killed himself. She was 13 years old when Hitler came to power, and in her worldview, Nazi society was normal and good. She lived in Nazi society like a fish lives in water. Hitler hired her when she was 22.

It wasn't until 20 years after the war, in the 1960's, that she began to comprehend that Hitler and Nazism were bad. Books and movies started coming out, revealing the horrors of the Third Reich. This forced her to confront the notion that the culture she considered to be good and normal was, in fact, sick and viciously destructive.

She was quite intelligent, and she had actually liked Hitler as a person. He was a warm, friendly man who had treated her with kindness. The secretarial group was a small domain that was isolated from Hitler's less genteel activities. At that time, the notion that Hitler was a murderous, diabolical monster was simply unimaginable to her.

How was it possible that she had misjudged him so completely? Why had she not seen the truth? She felt that she had been a bad person for not figuring out what was going on. But she was not alone. Millions of others also failed to discover the truth, or acknowledge it. For the rest of her life, guilt and shame fueled a painful struggle with mental illness.

Because her consciousness was expanded, Junge was forced to rewrite primary portions of her worldview, and this was quite traumatic. Life is never the same once you awaken to the reality that your culture is insane. Awakening liberates you. It allows you to think outside the box, and see with greater clarity. Eventually, near the end of her life, she learned to forgive herself. She wasn't a bad person — she was raised by a bad culture.

Junge remained silent about her time with Hitler for more than 50 years. A documentary of Junge's life was released shortly before her death in 2002. The title of the film was *Blind Spot*.

Ghost Dancers

In the 1860's, deranged Europeans rode their brand new railroads into the Great Plains and commenced the Buffalo Holocaust. By 1890 the buffalo were at the brink of extinction. For the Indians, the buffalo were the core of their existence. With the buffalo gone, their traditional hunting life became obsolete and impossible.

In January 1889 a Paiute prophet named Wovoka had a vision. He encouraged folks to perform the Ghost Dance. The people should gather and dance five days and four nights, then on the fifth day bathe in the river. It was powerful dancing, with up to 500 in a large circle. Many enjoyed manic frenzy, some fell into hypnotic trances, some collapsed and became unconscious. The Ghost Dancing spread from tribe to tribe, eventually expanding over a vast region of the American west.

The vision was that if the Ghost Dance was properly performed, a great flood would come and wash the whites back across the ocean to their European home, where they belonged. The ancestors and the buffalo would be reincarnated, the epidemics of white diseases would cease, and life would return to the good old days. It was a magnificent plan. Unfortunately, it was not an immediate success. No matter how much they danced, the whites didn't go away, and the buffalo did not return.

A different version of the Ghost Dance story, described by Lame Deer, used the metaphor of rolling up and disposing a filthy old carpet, which was a symbol for the white man's world. Imbedded in the filthy carpet were roads, mines, cities, farms, factories — every form of ugliness that civilization brought with it. When the old carpet was rolled away, beneath it was revealed a healthy wild natural Fairyland — the land as it was prior to the invasion of the whites and their diseases.

The Ghost Dance became so popular that the white invaders began to get spooked. They were intimidated by the fact that large numbers of Indians were enthusiastically dancing, with the goal of cleansing their lands of the white race forever. They perceived (incorrectly) that the natives were conspiring to launch a violent rebellion. The army was called, and several hundred Ghost Dancers were exterminated at Wounded Knee on December 29, 1890.

The Indians were screwed. Their buffalo hunting culture was over. Escape or migration was not practical. Driving the whites away via warfare was purely suicidal — they were hopelessly outnumbered and outgunned. Despite the odds, brave men like Geronimo, Crazy Horse, and Sitting Bull fought hard to defend the sacred motherland and preserve the American way of life. But they lost the war, they lost their home, and they lost their way of life.

The whites demanded that they become Christians and farmers, and assimilate into civilized society — a notion similar to expecting blacks to blend into the Ku Klux Klan (the heroic defeat of Custer at Little Bighorn triggered a tidal wave of racist hatred, even including most Eastern liberals). The only remaining option for the Indians was utilizing the spiritual power of their culture. Only miracles could save them now. So they tried. They had nothing to lose.

John Michael Greer is a thinker and writer who has some Lakota blood. He wrote that magic was an important part of Lakota culture. It helped their society run smoothly. Magic is a powerful technology, because of its ability to direct the collective imagination of a human community. The Ghost Dances were an attempt to use old strategies to confront a new challenge.

Greer says that this is a common scenario in history. When times change and a culture no longer works, people have a tendency to cling to their old culture even harder. He sees this being repeated today, as we confront the end of the era of cheap energy. We are furiously using the magic of our failed culture (science and technology) in an attempt to continue growing forever on a finite planet. Instead of taking intelligent action to sharply reduce consumption and population, we are circling our gas guzzlers and performing the Technology Fairy Dance, to bring back the gushing oil wells.

It is tragic to be possessed by a dysfunctional worldview. The civilizations of the Fertile Crescent destroyed their ecosystems and crashed. Then, the Greeks repeated the same mistakes, destroyed their ecosystem, and crashed. Then, the Romans repeated the same mistakes, destroyed their ecosystem, and crashed. Then this self-destructive way of life was exported to every corner of the planet. Today we're repeating the same mistakes all over the world, but in a way that's 1,000 times more destructive.

When we run out of oil and gas, we'll burn coal like crazy, and when we run out of coal, we'll burn the forests like crazy. We'll manufacture PV systems, windmills, and fuel cells like crazy — all in a futile effort to extend the life of industrial civilization for a few more years — to desperately cling to our automobiles, televisions, and air conditioners for just a few more years.

During World War II, most of Japan's oil imports were cut off. At this point the Japanese began converting their forests to *pine root* (turpentine) for use as a fuel in airplanes, trucks, and industry. Naturally, this resulted in substantial deforestation. In wartime Europe, hundreds of thousands of cars, trucks, and buses were powered by gasifying wood chips or charcoal. Today, wood-burning power plants are being built. Rainforests are being ravaged to make room for palm oil plantations (vehicle fuel) and soybean fields (livestock fuel).

What will we do as the fossil fuel era winds down? Will we pick up our axes and turn the entire planet into a treeless Easter Island? Will we obliterate the planet or grow up? Time will tell. I'm voting for growing up.

Ragnarök — Nature Bats Last

My Norse and Germanic ancestors told an ancient story about the mythical battle of Ragnarök, which seems to be eerily prescient of modern times. It describes the failure of powerful humanlike gods to subdue and control the forces of nature. It's a story of destruction and world renewal.

These Teutonic myths discuss two families of gods and goddesses, the Aesir and the Vanir. The Vanir were the peaceful and benevolent earth deities of the Old Europeans, possibly the original indigenous tribes of northern Europe. They made sure that the pastures and forests had adequate rain and sunshine. The Vanir had four primary deities. Njörd (the sea) and Nerthus (earth mother) had two children, Frey (summer sun and rains) and Freya (beauty, love, fecundity). When Frey was born, the gods gave him Alf-heim (Fairyland).

The Aesir family of gods and goddesses arrived later, with the invasion of Indo-European farming tribes from the southeast. They were the deities of a culture of warriors and conquerors, and they were accompanied by domesticated animal slaves — horses, goats, boars, and dogs.

The stories say that the Aesir conquered the realms of the Vanir, and then they fettered the world's chaotic natural forces — Surt the giant (volcanoes and earthquakes), the Midgard Serpent (turbulent seas), the Fenris wolf (powerful animal wildness), and Loki the trickster (fire and air). Thus the Aesir domesticators subdued nature, temporarily. The Aesir had foreknowledge that their control would not be permanent, that the system they created would eventually be destroyed, and that they would be killed in a great battle. Every civilization has an expiration date.

With nature fettered, the world became unbalanced. There was a string of three severe winters, wolves swallowed the sun and moon, the stars fell from the sky, earthquakes pulverized mighty mountains, and human society degenerated into an age of crime and terror, brother slaying brother. Finally, the fettered forces of wild nature broke free, and challenged the Aesir gods. This was the battle of Ragnarök — the twilight of the gods.

Almost all of the combatants on both sides perished in the battle. Then Surt the giant spread fire over the whole Earth, burning up everything, leaving behind nothing but naked soil (global warming?). The flames purified the Earth, and prepared the way for a new order in the world. Then the rivers and seas rose up, and all dry land sank beneath the surging waters of a huge flood.

In *Norse Mythology*, Peter Andreas Munch announced the dawn of the new era with a beautiful line: "Out of the sea there rises a new earth, green and fair, whose fields bear their increase without the sowing of seed." All evil was passed and gone. Two mortals survived Ragnarök: Lif (life) and her husband Lifthrasir (vitality). Their offspring brought forth a new and nobler race of humankind, who lived in never-ending joy.

Several assorted minor deities also survived, including two of Odin's sons, Vali and Vidar, who ruled the newly regenerated world. Vali was a god of light, personified by the lengthening days — the renewal of the year. His brother, known as Vidar the Silent, lived in Landvidi, a green place surrounded by the solitude of a vast and impenetrable forest. Vidar's pathless primeval forest had never known the sound of the ax or the voice of man. The story of Vidar shows us a love of wild nature that is powerful, absolute, and uncompromising.

Obviously, the story of Ragnarök was not the product of a peaceful and harmonious society. It's intriguing that their vision of a future paradise was one of wildness, freedom, and balance — a way of life very different from the reality of their turbulent era. It is a vision that has striking similarities to the vision of Wovoka and the Ghost Dancers — sweeping away the madness and returning to the good old days of living in harmony with nature. A thousand years ago, the mindsets of tribal Europeans and tribal Americans had much more in common. It is a dark sign of our times that we no longer remember our ancestors' dreams of pure wildness and freedom. We must remember then.

Most of what we know about the Teutonic myths comes from the poems written down in the 1220s by an Icelandic bard named Snorri Sturluson, a Christian who remained fond of his ancient indigenous culture. By this time, most of Europe was dominated by Christianity. Snorri wrote down the poems to preserve memories of an older culture that was deliberately being driven to extinction — in a manner similar to how the Americans tried to erase the culture of the Indians — compulsory assimilation.

Benefits of a Shrinking Herd

Today, the global production of energy is close to peak. Following the peak of energy production (or sooner) will come the peak of food production. Following the peak of food production (or sooner) will come the peak of human population — at last! These peaks will be followed by a steady decline in energy production, and therefore food production, and therefore population.

This is a beneficial outcome for the ecosystem, because it means that our population is going to go down, no matter what we do, and it will require no effort whatsoever on our behalf. We won't have to talk about it, we won't have to pass and enforce strict laws, and we won't have to suffer through endless hysterical arguments about human rights, eugenics, racism, imperialism, the will of God, or whatever. The downsizing will be automatic, and impossible for us to prevent. But we will also have the option of accelerating the healing process by voluntarily limiting our own breeding, and by making other positive lifestyle choices.

With fewer people, and fewer machines, the destruction of the planet will be reduced. Our ability to mine soils, forests, fish, water, and minerals will be reduced. Our global transportation system will fall apart, which will terminate the global economy as we know it. As human society shrinks in scale, our wild relatives will finally be able to begin the process of recovery — cod, salmon, grizzly bears, and buffalo. With time, the air will get cleaner, and so will our lakes, streams, and oceans.

The rules of the game will change. More and more options will become available to us as the herd shrinks from seven billion to four billion, to one billion… Significant population reduction will provide us with delicious opportunities to make fresh starts.

We can quit irrigating deserts, and emptying aquifers. We can pull down the fences and let the buffalo return to their ancient home. We can turn cornfields into grasslands and agroforests. We can celebrate the recovery of the world's wild fish. More and more beautiful things will become possible as Mother Nature (or human intelligence) guides the human herd back to a place of health and balance. Before long, it will become possible for us to quit the practice of enslaving plants and animals. But, will we be able to deliberately abandon our bad habits by a willful choice? A crucial question!

The transition from a machine-powered society to a muscle-powered society will be automatic and effortless, because we will eventually run low on the energy needed to make and use machines. This transition will turn back the clock 200 years or so — but it won't save the world, and it won't make human society sustainable. Two hundred years ago, most humans were the inhabitants of unsustainable civilizations.

So, the good news is that the reduction of the human herd will be automatic and effortless. The more challenging news is that the return to sustainable living — wild, free, and happy — will not be automatic and effortless. It will require that we succeed at doing something really, really hard. It will require that we break out of our 10,000 year cycle of repeated mistakes. In order to accomplish this, we will have to remember or reinvent belief systems that are healthy, wholesome, and intelligent.

Humans are a stumbling dance of emotion and reason. Emotion seems to be humankind's dominant decision-maker. We got to where we are today by a long process that began with dazzling cleverness, but went sideways and snowballed into a colossal worldwide disaster. Consumer society is like a long-term hardcore meth user. The meth was available, we had the money, and it started out being a lot of fun. But now our teeth are gone, our hair is falling out, and our skin is covered with oozing sores. With an imbecilic grin, foolishness shouts "full speed ahead!" But intelligence insists that we completely break out of this suicidal way of living, and use our immense cleverness to find the path to healing. Only a new and radically different form of cleverness can reverse the evils that cleverness begat in ages past, and it must constantly outwit its faithful and powerful shadow, foolishness.

Danny Billie, a Traditional Seminole, once said: "You're trying to say that you can live in the modern way and continue to think in the traditional way. That's not true. The way you live affects the way you think." And, of course, the opposite is also true: you can not live in the traditional way and continue to think in the modern way.

A happy sustainable future will not be possible until we turn our worldview inside out, and return to a mode of thinking and living that includes an intimate sense of connection to the land, and a profound sense of reverence and respect for life. This transition is now underway, but the complete process could take several centuries.

The healing of our worldview will be given a tremendous boost by the eventual collapse of the consumer way of life. We consumers suffer from stressful lonely lives in an unpleasant human-built reality that isolates us from the rest of the natural world. We spend just a tiny amount of our lives outdoors. We buy our food, clothes, and tools from stores. We don't know how to make most of the things that we possess, and most of what we possess is frivolous and unnecessary.

As the collapse unfolds, we will have no choice but to move toward a way of life that is much healthier than how we live now. To a large degree, our human-built world is going to rust, rot, crumble, and return to the Earth. Once our TVs, cell phones, and computers go black, and we are forced to go out into the living world and gather acorns and dandelions for dinner, then our worldview is certain to change in positive ways. We will develop an intimate and far more respectful relationship with the land where we dwell. We will come to appreciate the wisdom of the chipmunk, coyote, and raven, who have never forgotten how to live, and live well. We will discover stars, birdsong, and the perfection of creation. It will be a beautiful homecoming for the long-lost consumer space aliens.

Our return to the sacred family of life will also lead to a *radical* change in our thinking, because we will be making a *radical* shift in the way we live — from industrial consumers to foraging horticulturists and herders, or something. We won't be merely sewing patches on a worn out antique worldview. We'll be moving in a bold new direction, toward health and balance.

Another benefit of the collapse is that functional communities will become the norm once again, because loners will have less success at survival in a world without grocery stores, police, and cheap energy. Modern consumers often suffer from painful loneliness, despite living amidst vast hordes of humans. We isolate ourselves by embracing distractions — television, radio, internet, music players, video games, books, pets. We suffer in space alien neighborhoods where the residents come and go via automobiles, and have little contact with one another.

There will come a day when we remember and resuscitate good old fashioned community — caring, sharing, cooperation, celebration. Once again we will live our lives in a community where seeing a stranger would be a rare and peculiar event. The curse of high mobility will be lifted when motor vehicles finally go extinct. We will be freed from the chains of our unhealthy individualism — and this will result in a beautiful healing process for the lucky descendants of civilized humans.

So, collapse will change the way we live and think. It will shrink the human herd, which will open the door to new possibilities. We will finally have a realistic option of giving up our bad habits, and returning to a wild, free, and happy way of life. This won't be automatic and effortless, and it may be our final opportunity to extend humankind's existence on the planet.

We can begin pursuing a happy sustainable future today. We don't have to wait for the collapse to go through all of its storms. Worldviews are not carved on stone tablets. Worldviews are software, and software is editable. All progressive movements first began in the hearts of individuals. Creative people can play an important role in how society thinks. This may be the greatest art project in human history: presenting visions of a sustainable tomorrow — imagining a safe return to our ancient home, Fairyland.

We need to learn and grow, and come to see things from a new perspective — a perspective where humans are not superior, where humans are sacred guests in a sacred land, where we will be welcome to stay only if we speak, think, and behave with complete respect for all beings. A whole lot of the work that needs to be done now — really, really important work — is inner work. Learning is healing. Questioning is healing. There is also much important outer work to be done in our communities — follow where your heart leads, and use your gifts.

Reevaluating Necessities

In his book *Inside Africa*, John Gunther wrote that all British people in colonial Kenya religiously wore sun helmets whenever they were outdoors. They religiously taught their children that if they ever went outdoors without a sun helmet, the sun would promptly kill them. Everyone accepted this fact as a certain, unquestionable truth. Then came World War II, and American soldiers arrived in Kenya. The Americans never wore sun helmets, and none of them were ever killed by the sun — even after months of exposure. This was an absolutely mind-blowing experience for the Brits. Before long, they quit wearing helmets.

In consumer society, we have many sun helmets, but the one with the biggest impact is the automobile. Consumers are terrified of going anywhere without an automobile. They would be exposed to the weather (and might muss their hair). They would be vulnerable to the savage criminal hordes. They would be committing treason against their class. The neighbors would start talking about them (did they lose their license for drunk driving?). Fortunately, a growing number of consumers are discovering that moving around without an automobile isn't fatal — it's healthy, thrifty, virtuous, and far more pleasant.

Most things that we consider to be necessities are not. In my lifetime, many new products have been added to the list of ersatz necessities: televisions, computers, cell phones, microwave ovens, cable TV, digital music players, etc. The advertising industry works hard to brainwash us into believing in ersatz necessities. It is really easy to get swept away by it all, and really expensive, in many ways.

Annie Leonard is a first class eco-preacher. In her video documentary, *The Story of Stuff*, she points out an uncomfortable truth: folks in developed countries have ceased being *people*. Following World War II, we've been transformed into a strange new life form called *consumers*. "Consume" means to use up or destroy. Consumers are not the slightest bit embarrassed by this insulting label.

We live in a society with a consumer worldview. Consumption is essentially our dominant religion, our purpose in life, our identity. Quite literally, we have become slaves to fashion. Shopping is as vital as breathing, we can't live without it. Our value as a person is judged by what we consume, and how much we consume. Consumers are typically driven by three desires: to have more stuff than they had last year, to have more stuff than their parents did, and to have more stuff than their neighbors do.

Diabolical minds in the multi-billion dollar pet industry are working hard to delude us into believing that pets are goofy consumers, too — and they're succeeding. A growing number of pet owners have become convinced that their animals literally *expect* and *deserve* a first class consumer lifestyle (including a daily dose of Prozac). Pets have a God-given right to relax in pet resorts, wear expensive costumes, devour gourmet feasts, and enjoy weekly massages. To deny them a life of luxury and excess is cruel.

My great-grandparents, Anton and Pauline Schneider, raised six children in Pittsburgh. Their house was on a hill on the south side of town. Access was by a footpath from the road below. Municipal water was provided by a water spigot shared by several homes. Children would carry water home in buckets. Each family had an outhouse. They had no car, no phone, no electricity, no computer, no refrigerator, no television, no radio, no washing machine. My great-grandparents in North Dakota lived in sod huts for a while, as did their neighbors.

This simplicity was not a problem or a handicap. The whole community lived like this. In many ways, their lives were richer, because they enjoyed far more interaction with their family and community.

I will always remember a trip I took one summer along the Lake Huron coast of Ontario. I drove through an Indian reservation around sunset. In the back yards of many homes, campfires were burning. People were gathered around the fires to sing and talk — to be together, and to enjoy each others' company. It was so healthy! The experience stuck some deep ancestral chords.

Never forget that the lifestyle of modern consumers is extreme. It's hard to comprehend this, because consuming is the only way of life that we've ever known. We actually think of it as being normal and acceptable. To put our lifestyle into perspective, consider the following:

- Around the world, close to a billion humans suffer from hunger.
- A billion or two families do not have cars.
- About 1.6 billion people have no electricity.

- About 2.5 billion people burn wood, dung, or biomass for heating and cooking.

- In 2001, when there were 6.1 billion people in the world, 2.7 billion lived on less than $2 per day.

- Most homes in India do not have flush toilets — and this is true for many other countries, too.

Of every seven trees cut down, one is turned into toilet paper. Billions of people do not use toilet paper — a simple splash of water will do the trick. But the industry is growing rapidly, as people in developing countries are being brainwashed into believing that toilet paper is an essential component of the modern, hygienic, civilized way of life.

It's the same story with disposable diapers. Diaper sales are skyrocketing in China. In earlier decades, Chinese infants were dressed in comfortable open-crotch pants. When nature called, they simply squatted and let it fly. This worked well in rural areas, where people spent much of their time outdoors. Many Chinese kids now live in cities, doomed to an indoor way of life that worships trendy convenience at any cost. More and more of these kids are discovering the pleasures of diaper rash.

Chinese people in rural areas often don't heat their homes during the cool months. Instead, they dress warmly (a practice that consumers will remember some day). High-tech heating and cooling systems are not necessary.

The collapse of the global economy is going to be a great equalizer. Consumers are moving toward an era of *involuntary* simplicity, a smarter, healthier, and far less wasteful mode where our lifestyles will not be much different from the people who pick our coffee beans. In this process, we will learn many important things, and free ourselves from countless illusions and bad habits. A vital notion to understand is that *standard of living* is not closely related to *quality of life*.

Thirty years ago, a friend of mine travelled in Thailand. The experience stunned him. The Thais were the happiest people he had ever met, yet they owned almost nothing. When he returned to Kalamazoo, and looked with disgust at his huge collection of stuff, he hauled two-thirds of it out to the sidewalk and gave it away. That made him feel better.

I got rid of my ancient car two years ago, and I really haven't missed it (I occasionally ride my motor scooter). I hired a taxi once, when I had stuff to haul. That ride cost $12. The town where I live is somewhat bicycle friendly, so I mostly bike. It's far more relaxing to leisurely pedal down the alleys. Car drivers tend to be stressed, irritable, anxious, distracted, always in a big rush. It's pleasant to avoid that icky craziness, get healthy exercise, and save a bunch of money at the same time.

The compulsion to hoard ersatz necessities and frivolous impulse purchases condemns us to a life of endless drudgery at the workplace — endlessly making excess money in order to acquire a growing hoard of unnecessary stuff in order to gain acceptance and approval from a peer group that has been culturally programmed to value high-waste living.

The fast, easy, and intelligent path to wealth and prosperity is to have few wants. When you mindfully limit your consumption, the need for money-making declines. This liberates precious time that can be used for doing things that you find to be interesting, meaningful, and rewarding.

Joe Bageant was a sparkling critic of modern America, where everyone believes that they have a divine right to live in a huge McMansion, with a swimming pool, a big flat screen TV, and three cool cars in the garage. He wrote that Americans have come to perceive that a cell phone has more inherent value than a dozen eggs. None of our ancestors owned cell phones, but all of them ate food — every day. At the bottom line of species economics, the only real currency is the calorie. You can't eat gold. There will come a day when chickens and firewood are understood to be far more valuable than giant Chevy pickups.

Reevaluating necessities is powerful medicine. We waste so much. Living lighter generates a lot of pride and satisfaction, and it makes a positive contribution to healing our worldview. Living lighter is not against the law, we can start right now, and get better every year. If we choose this path, we're more likely to meet and befriend allies on the path to growth and healing.

We can speed the global healing process by dragging our feet, withholding support, rocking the boat, being uncooperative, questioning everything, inspiring doubt, mocking the glorified lords and ladies, and strewing the path of industrial society with banana peels. Why should we waste our lives striving to earn the respect and admiration of the daffy stuff-hoarding hordes?

I make a deliberate effort to not think or behave in a manner that consumer society deems proper and appropriate. Nobody can force me to buy a McMansion. Nobody can force me to drive a car. Nobody can force me to spend my leisure time indoors, staring at a glowing screen. Nobody can force me to create children or enslave animals. And so on.

Today, it is virtually impossible to live in a sustainable manner, but it is totally possible to live far less destructively than the herd. And it feels good. Work less, earn less, spend less, and have more fun. There is big truth in the old cliché that the best things in life are free. Seek them.

Healing Ourselves

A while back, a friend spent time with the Micmac teacher, Albert Ward. She was furious and frustrated about how we civilized humans were destroying the planet. Albert told her that if she wanted to heal Mother Earth — which he really felt didn't need our help — the first step was to heal herself. She needed to discover who she was, and then she needed to be herself.

This notion blind-sided me — that the Earth is completely capable of healing itself. We're so programmed to believe that controlling the

Earth is the human's purpose in life — our sacred obligation as the "superior" species. We can barely control ourselves.

I now know that this is a common belief in many Native American circles. They have absolute faith that industrial society is going to continue on its present course and fly off the edge of the cliff. In the end, Mother Nature will restore balance, one way or another — this is the only possible outcome.

But what did Albert mean when he said that she needed to discover herself and be herself? That's a splendid subject for a rich voyage of contemplation! Who is the wise and ancient human soul that dwells deep within us?

In his book *No Word for Time*, Evan T. Pritchard wrote that animals do not suffer from lost identity problems. A raccoon knows how to be a raccoon. A tiger knows how to be a tiger. But civilized people have forgotten how to be humans. We've forgotten who we are, how to live, our place in the world. One of the most fundamental things we need to remember is what it means to be a genuine wild and free human — something that we've been taught to look down on.

Imagine walking in a deep forest, passing through a magical veil, and coming face to face with one of your wild and free ancestors who lived long ago. Look her in the eyes. Feel her presence. Sense her health and vitality, her strength and solidity, her awareness, the sharpness of her senses. Try to imagine her inner world, her perception of reality, her values and beliefs, her ancient culture of cooperation, sharing, and equality. She is your flesh and blood. You are her living descendant. You carry her genes. Yet there is a tremendous gulf between you — two very separate realities, having little in common, like a wolf and a dog.

Can you remember what it feels like to be free? Imagine living in a world with no presidents, police, laws, or private property. A world without walls, fences, roads, and bridges. A world with no keys, no locks, no machines, no guns, no jobs, no bosses, no money. A healthy world of abundant life. Forests filled with singing birds, lakes and streams filled with fish, prairies filled with deer and buffalo. Can you imagine feeling at home in the natural world? Can you remember what it feels like to be a human being? Doesn't it feel good?

In his wonderful book, *Original Wisdom*, author Robert Wolff described the years he spent with indigenous people, gathering information on their healing arts, in an effort to preserve the knowledge before it was permanently erased by progress. Wolff spent time among the Sng'oi tribe of Malaysia. A Sng'oi healer took him on long walks in the forest, and helped him learn how to turn off his thinker and discover the power of his ancient wild mind.

Wolff learned how to blend into the forest and become one with its powerful harmony — this made him feel incredibly alive. He learned how to sense the presence of others — even if they were miles away, he intuitively knew where to find them. He learned how to communicate without words, in a manner he described as telepathic. His Sng'oi teacher sometimes knew about events before they actually happened.

Wolff learned the skills of herbal healing via a sense of *inner knowing*. When he had a headache, he saw a vision of the plant that would help him, and how to prepare it. When he did just as the vision instructed, his headache was promptly cured. Many scientists believe that primitive cultures learned herbal healing via trial and error, but there are ancient ways of knowledge that scientists have yet to learn about. Many skilled herbalists say that they can comprehend the plants communicating with them.

With years of practice, Wolff came to remember the important notion that being human means being spiritually connected with the natural world. He became a stronger man, and a contented man. But there were times when he wished he could turn off his power of inner knowing. In urban areas, surrounded by crowds of strangers, he would feel overwhelmed by sensing the mental activity of those around him —

many swirled in feelings of anger and resentment, and many seemed to be disconnected from their feelings. It was awful.

People who have access to their indigenous soul can live in a wild land and be at home with other species. They can form strong relationships, engage in non-verbal communication, and build respectful bonds.

In his book *Never Cry Wolf*, Farley Mowat described the indigenous people of the Canadian north. He told tales of Ootek, a man who understood the language of wolves. One day, Ootek heard the wolves speaking about humans moving across the land — a few hours later three human visitors arrived in the camp. Another time, the wolves spoke of caribou moving through the region. Ootek quickly packed up for the hunt. He discovered caribou exactly where the wolves said they were. Like the jungle drums of Africa, one wolf announced the news, which was picked up by another wolf a mile away, and rebroadcast to others, who relayed the news further out — Ootek didn't get the news directly from the source, but from a wild canine communication network.

It is so important to remember the power of inner knowing, the ways of our non-human relatives, the realization of the oneness of all life, the joy of being completely alive, the sensation of feeling the love and power of the universe pulsing through you. These things are core components of being fully human — of being whole and healthy.

It's so important to spend generous amounts of time with the spirits of our wild and free ancestors — to come to remember them, know them, understand them, like them, care about them — to spend the rest of our days moving back in their direction, moving back home, where we belong — safe, secure, at peace — wild, free, and happy.

The Prison of "Positive Thinking"

In her fascinating book, *Bright-Sided*, Barbara Ehrenreich delivered a powerful dope-slap to the trendy cult of positive thinking. When

Ehrenreich got breast cancer, she came into contact with well-intended people who tried to encourage her with notions like: cancer is a positive, life-changing experience — a gift. It will make you prettier and stronger. If you die, it's your own fault, for slipping into negativity (blame the victim). This lunacy absolutely infuriated her. She was certain that her cancerous breast was the result of living in a toxic society, not negative thinking.

America is the fatherland of positive thinking. Charismatic gurus in the self-help industry are raking in a fortune by convincing the feeble-minded that positive thinking is the gateway to health, wealth, and prosperity. Believers are encouraged to deliberately suppress any and all negative thoughts, lest they get cancer and die. Negative thinking is the mother of all misfortune in your life, they say. Negative people suck! Throw all negative people out of your life. And so on.

The latest trends in corporate management also embrace positive thinking. Ehrenreich suggests that positive thinking was a driving force behind the financial meltdown of 2008. People in the finance industry who could see huge trouble brewing just put on a positive smile and kept their mouths shut. Those who could not restrain themselves, and openly expressed their reality-based concerns, were immediately fired. In the modern corporate world, "complainers" are deliberately weeded out, to reduce friction in the organization, and to permit the enterprise to soar away on flights of pure fantasy. If you discover a fire in the boardroom, keep quiet and think positive thoughts.

With regard to the Earth Crisis, positive thinking is an absolute nightmare. A positive thinking consumer is going to wish for miraculous technological solutions to all of our myriad problems (and for a new SUV, a bigger house, a better job, and losing 20 pounds) — solutions that require no lifestyle "sacrifices" at all. They are wishing to extend the survival of the most destructive way of life in human history. Instead of accepting responsibility for the way they live, and living more mindfully, they indulge in positive thinking, and trust that everything will magically get better, somehow. Maybe a fairy godmother will fix things. Keep smiling.

At the same time, they are disregarding big warnings. People expressing important and legitimate concerns are perceived as being negative people — the dreaded "doomers" who want to spoil our delightful tea party. Away with them! They aren't fun to play with.

The mother of all questions is: what is genuinely *positive*? Is it really *positive* to wish for the extended survival of our industrial civilization, despite its terrible costs? That seems like wishing that World War III will never end (pay no attention to the miserable millions behind the curtain), because it's our most convenient option for now, thank you very much. Change is a hassle, we don't have time, we aren't interested.

I believe that it's positive to dream of healing, to imagine a sustainable tomorrow. It's important to smash down the walls of our mental prisons and become present in reality, in the fullness of the darkness. It's healthy to acknowledge the existence of problems, and respond to them in an appropriate manner. I remain hopeful for the long-term future. Human foolishness must be finite, right?

Another name for robotic positive thinking is *blind optimism*. Sufferers of blind optimism wishfully bet everything on good luck. At the other extreme, the traditional pessimist is a gloomy being who robotically expects the worst, today and forever. Neither of these types are notable for working hard to achieve beneficial goals.

In her book, *The Positive Power of Negative Thinking*, Dr. Julie K. Norem described a non-robotic third option: *defensive pessimism*. Defensive pessimists live with their eyes open, and anticipate that everything may not go as planned. Like a bicyclist riding in heavy traffic, they scan the horizon for potential problems, and contemplate how to respond to them. They are better prepared to respond effectively to challenges when they appear. You might even call them realists.

I am a defensive pessimist. I can imagine possible problems on the road ahead. The Technology Fairy may fail to miraculously save us. We may get clobbered by peak energy, peak food, peak water, etc. What are our best options? What would our wild ancestors recommend? Could defensive pessimism actually be more intelligent than deliberately suppressing any and all "negative" thoughts?

Clear Thinking & Deep Awareness

Walter Bresette often told a story about the four instructions that the Creator gave to the Anishinabe. He told it with a gentle voice and a warm smile. The bare essentials of this story are thus (Walter presented a much longer version):

> Before the beginning, there was a thought and a vision. The Creator called all of the people to the fire. And at that fire, the Creator gave instructions to each of the peoples. And he told us, if you ever forget your instructions, all you have to turn around and look back. That's all you have to do is turn around and look back at the fire — and you'll remember your instructions.
>
> We have four commandments, to use a term that people will understand, for staying on the red road.
>
> The first direction is represented by sweet grass. Always be gentle.
>
> The second direction is represented by that grandfather tree over there. It stands tall, and that represents truth. Stand up, be tall, speak the truth.
>
> The third direction is represented by the spirit of the deer — sharing. The spirit of the deer shares, gives, provides nutrients. We take the deer, but the commandment is to give, to share.
>
> The fourth direction is represented by the oldest of the grandfathers, the stones — the stones of the sweat lodge —

the ancient wisdom of the very Earth itself, and that is to be strong.

If we are gentle, tell the truth, share, and are strong, then we can stay on the sweet grass road. And you know something — that's the hardest damn thing to do! Four simple things, four simple instructions, and we would be so happy.

The Anishinabe only require four instructions, because they have an ancient and functional culture. Modern consumers need to add a fifth instruction, in their search for the sweet grass road: *be present in reality*.

The consumer world bombards us with endless distractions, the entertainment industries never sleep. We have access to a ghastly cornucopia of titillating fantasy worlds. Indeed, it's quite easy to spend most of our lives absent from reality, bouncing mindlessly from one distraction to another — and many do. Distractions provide the illusion of an escape from reality. Life in the 21st century is more than a little painful, and even the most amazing fantasy voyages cannot rinse this pain away and give us inner peace, but inner peace is really what we are starving for.

Needless to say, the millions of people who strive to remain absent from reality are not going to be available to devote their life and love to the healing process. We can only contribute to the healing process when we are present in reality, with both eyes wide open, and a heart filled with love.

Yes, being present in reality exposes us to the fullness of the darkness, which hurts. But I can solemnly testify that this pain will not kill you, it makes you stronger. Being present in reality also exposes us to the fullness of the beauty, which does not hurt at all. The living world is far from dead, and it provides abundant spiritual treasure for those who seek it.

Michael Ventura co-authored a book titled *We've Had a Hundred Years of Psychotherapy — and the World's Getting Worse*. He says that we are living in an avalanche that nothing can stop. Many worry that the world is coming to an end, but that's not true. The only thing that's dying is our crazy civilization. There is no safe and secure place in an avalanche, so don't bother looking for it. When you live in a time of decline, you do the work of the soul. You try to be a wide-awake human and keep alive what you think is beautiful and important, because some day the Dark Age will end.

Joanna Macy is a Buddhist philosopher and healer. She believes that our denial and apathy are rooted in our fear of pain. The world is hurting. She recommends overcoming that fear, and acknowledging our pain. The other side of that pain is our love for the world, our connectedness with all life. It's OK to grieve and mourn. It's OK not to be optimistic. The main thing is simply to be present, to keep showing up and loving the world.

John Trudell, the Santee Dakota visionary, actually came up with brilliant and workable solution for all of humankind's problems. Here it is: *clear and coherent thinking*.

Clear thinking allows us to accurately perceive and understand the reality in which we live. The enemy of thinking is belief, because beliefs build mental walls that limit the freedom and scope of our thinking. When we become imprisoned by beliefs, we believe that believing is thinking — and this really inhibits the power of our intelligence.

Believing is easy, but thinking without blinders requires robust mental power and clarity. Clear thinking requires that no ideas are sacred, nothing can be off-limits to questioning. Our minds must be absolutely free. We cannot create a sustainable way of life without clear thinking.

Trudell used to annoy me. He is a brilliant man and an exceptionally clear thinker. I knew that he had certainly devoted tremendous clear thinking to this global predicament. So, what was he thinking? How do we get ourselves out of this mess? Give us the answers! *The answer is clear thinking.* Give us solutions! *The solution is clear thinking.* Of course, he's right. Sadly, our society has conditioned us to expect others to lead us and give us instructions. We are not a society that celebrates and promotes clear thinking. It's a skill we need to learn.

What is so tragic is that newborn humans are pure, unspoiled, wild beings. They are completely capable of living and thriving in a hunter-gatherer society — wild, free, and happy. They only turn into consumers by being born in a society having a consumer worldview, where they are parented and educated by consumers. Before long, their wildness is lost and forgotten.

Looking inward is essential, and so is looking outward, into the natural world. In his book *Last Child in the Woods*, Richard Louv proposes the notion that our younger generation is especially suffering from what he calls *nature deficit disorder*. More and more kids are spending their lives indoors, fixated on glowing electronic screens, rather than playing in the woods. This trend is not healthy, and it is causing tremendous damage, in many ways.

Tom Brown's book *The Tracker* introduces us to Tom's fascinating world of nature awareness, wilderness survival, and the skills of hunter-gatherers. His books provide powerful medicine for nature deficit disorder. Our healing process requires a homecoming, a full return to the natural world.

HOMECOMING

Marriage to the Land

My nine years in the woods of Upper Michigan was a time of studying, thinking, and searching. The information that I was absorbing was unpleasant to process, overwhelming at times. Throughout these years of learning and struggling, nature was my salvation, my refuge. I spent a lot of time walking, bicycling, picking berries, and hunting for rocks.

Nature was my medicine. Whenever my mind got too fried I would go outdoors. Being outdoors stilled my aching thinker, and provided a state of calm that allowed deep intuitive insights to rise to the surface. Nature is so sane and soothing. There is no madness in a quiet forest. I often experienced feelings of profound peace, while walking in the trees. Nature was alive and healthy and balanced. There were few inner demons that could not be exorcised by walking for an hour or so — and the return trip was almost always a voyage of euphoria and well-being.

The company of my wild relatives was delightful — the ravens, birch trees, bears, chickadees. They were easy to get along with, and they taught me many things. The tiny chickadees would be singing joyfully in the trees on bitterly cold days, when the grouse had buried themselves in the snow to keep warm. This was an important process of sacred remembering: reconnecting with the powerful harmony of the living land brought contentment and peace. I remembered glimpses of how life used to be in the days before civilization. It was good!

My meeting Walter Bresette was a huge experience — to spend a few days with someone from a far healthier culture, a culture that had not forgotten reverence for Creation. It was so healing to watch him hold a copper rock as a sacred object, as a being that was spiritually alive, and spiritually powerful. It was so healing to see someone who was passionately in love with the Earth — a rare pleasure.

Many years ago, I got into an online discussion with a Shoshone man. I had been foolishly ranting about "Native American spirituality," as if it was a standardized system of beliefs. He told me that there was no such thing. There were many tribes in the Americas, and each of them had their own specific and unique form of spirituality. Every bioregion was unique, and every unique tribe had a unique spiritual relationship with their unique place — a marriage to their land.

This connection, this sacred path of mindfully and spiritually inhabiting a specific place, had many facets. I once heard a Caribbean childbirth tale. They would take a newborn's placenta, wrap it around a coconut, and plant it. The resulting tree would be a lasting monument to the marriage between this clan, this place, and this new person.

The Shawnee had *naming trees*. At birth, a lock of the infant's hair was snipped, and placed in a fork of the naming tree. As the tree grew, the hair would be absorbed into the flesh of the tree. This tree became a powerful symbol of the community — the connection between the human clan and the green ones. Gray-haired elders could return to their naming tree, and reconnect to the events of their infancy — and their parents and grandparents. These trees were landmarks for people's lives.

People were also connected to the land by burials. The spirits and remains of their ancestors were all present in the land that they inhabited. The ancestral spirits were invoked to be present at the ceremonies. The spirits usually played a guardian role in the community — watching over their descendents, and helping them to successfully move through life. They were also sought out for counsel, when people were challenged by difficult problems.

In some places, the dead were buried beneath the floor of the dwelling, so they would be close to family affairs. In other places, they were buried toe-to-toe in a circle — similar to the manner in which they had sat around the fire while alive. Some people buried their dead naked, in a curled fetal position, like a seed waiting to be reborn. Some practiced sky burial, where fresh corpses were fed to the birds — so the sight of birds, and the sound of their music, always brought back memories of loved ones.

Wild people had an encyclopedic knowledge about their land. They knew every rock, stream, tree, and pond. They knew all of the plants and animals. They knew the medicinal uses of local plants. They could follow the tracks a field mouse across a dry gravel bed. They could tell which week of the year it was by the smells in the air — the buds, the blooms, the fruits, the decay. They knew their land as well as they knew their own body.

Some could "feel" the presence of an invisible deer sleeping on the far side of a hill. Or, the presence of a beached whale, miles away. Or, the approach of a group of humans, their location, the number in their party, while still many hours away. They knew how to communicate with the spirits of the ancestors. They knew the languages of animals, and how to converse with them.

Wild people were deeply rooted to their home. They celebrated their land, feasted on its beauty, cared for it, loved it, gave thanks to it. It was a deeply sacred relationship, their connection to the land. They knew that injuring the land was injuring themselves. They had a sense of belonging, which gave their lives purpose and meaning. They knew who they were, why they were here, where they were going.

Long ago I read an article about condors. It said that a free and wild condor soaring above the mountains was sacred, majestic, perfect. But a condor held captive in a zoo was less — far less. The essay concluded that *condorness* consisted of 10% condor and 90% *place* — hillsides, forests, crashing streams, big blue skies — the creature's birthplace, home, and temple. The same thing is true for humans, too. Humans who spend most of their lives indoors experience a tremendous and tragic loss of power and spirit and meaning. Only in Fairyland can we be fully human. All creatures are diminished, damaged, or destroyed by living in cities.

We can be fully human when we have a home, a place where we belong, a sacred connection to the planet. Chief Seattle was mystified by the fact that Europeans could wander so far from where their ancestors were buried and have no regrets. To his tribe, the homeland was an essential component of their identity. Leaving home would result in a loss of self.

In my nine years on the ranch, I tried in many ways to truly inhabit this land. I learned a lot. I had a map in my head that marked every blueberry nation, every apple orchard, every cherry tree. I knew the rocks well — every old mine site had rocks with unique colors, textures, and forms. I knew the roads and the trails and the streams. I spent much of my nine years wandering across the land — exploring, feasting, celebrating.

At the same time, I felt like I was only scratching the surface with my relationship to the land. I had been raised by a civilized culture. I was gray haired, yet knew almost nothing about the living relatives who inhabited this land — the names of the trees, or the birds, or the wildflowers. I was ignorant of many edible plants, and of most medicinal ones. I was not indigenous to this land. I had no elders, no guides, no teachers. I was an alien, an invader, a colonist.

Marriage to the land requires spending lots of time being outdoors, on the land, with your heart turned on — being reverently engaged in an intimate sacred relationship, celebrating the living holiness, and giving thanks.

Land of the Copper People

November 17, 2000. I am out of bed — relaxed, refreshed, renewed. It's 50° in the kitchen now, and I'm bundled up in two shirts and a jacket, waiting for the wood stove to heat up, and take off the chill. Winter is the season of soup. My huge Dutch oven sits on the wood stove, filled with beans, onions, dried carrots, celery, chilies, and smoked ham hock. The air is thick with warm and pleasing aromas. Mmm! Smell the soup! Yesterday was the first soup of the season. A celebration of the coming winter. A season of rest and calm and healing. Quiet and nurturing. A time for prayer and dreaming.

The sky is filled with clouds — a solid layer, medium thick, a dozen shades of gray. Outside my window is a land clothed white and pure with snow. Surrounding the house is a narrow strip of lawn that I keep mowed. At the far edge of the lawn are piles of Copper People, now dreaming beneath the snow. Then there is a band of the Grass People, golden stalks waving slightly in the gentle breeze. Beyond that is a band of brown woody brush, mostly alder. In the brush are many young apple trees, a few mature ones, and a worn and twisted grandmother apple tree, surrounded by a clan of her sapling children.

Beyond that is the beaver pond. The pond is the mother of life, the hub of the wild community, a popular drinking establishment, a gathering place. After the snow melt, thousands of tiny spring peepers roar with their love music. Migrating geese, ducks, and herons come to visit. The eagles, hawks, and owls soar overhead in search of a hot furry lunch.

The beaver pond is a temple of love. In late spring, the woodcocks are amazing — what a love dance they perform! They dance when the sunset glow is growing old, fading from fiery orange, to brilliant red, to the cool and gentle violet that ushers in the grays of the coming night. As the twilight deepens, the male woodcocks take wing and fly. Higher and higher and higher they fly, until you can barely see them. Their song is loud, ghoulish, rapid chatter of ascending hoots — eerie and ethereal — a most magical music of springtime nights.

When the male woodcock is high, high, high above the pond, he stops, falls like a rock, racing to Earth, pulling out of his dive at the very last second. And shivers of excitement and burning lust race up and down the female's spine.

A light shower of tiny hard snow pellets is falling now. Beyond the beaver pond is a band of forest. Maples, birch, hemlock, cedar. The evergreens are wearing a white mantle of hard and crusty snow on their boughs. Beyond and above the forest are the tall poor rock piles of the Franklin Mine, a mile to the west.

Let's climb to the top of the poor rock pile. The stones are a dark gray basalt, studded with pellets of greenstone, and splashes of white calcite crystals (the stuff that pearls are made of). And for the cunning hermit — who can still his mind, lose all thought, and concentrate his power of observation — there is treasure to be found. Sacred rocks.

Among the sacred rocks are the Copper People, ancient beings of power and wisdom. Red metal. But on the piles, they are bright green, turquoise, black, gray, or brown. The treasure hunter looks not only for colors of copper, but also for the unique shapes. The Copper People are knobby, gnarly, winged, protruding. You can know them by blind touch alone, because of their dense weight. You can also identify them by the rough texture of their skin.

The Silver People are much more rare. Sometimes you find pure nuggets, but more often they form in patches and crystals on copper hunks, in stones that the locals call "half breeds." On the piles, weathered silver looks the same as weathered copper — lumpy, dark, and dirty. When brushed, though, the Silver People shine white and brilliant.

Rarer still are the Datolite People — a hard people — gemstone hard. Local jewelers cut and polish them. And the colors, what a delight! Pink, maroon, tan, gray, yellow, but most often a pure and solid white, sometimes adorned with a darker spray of fine mossy lines.

Back to the peak of the poor rock piles. They sit on top of the spine of the Keweenaw Peninsula, which is actually an island. The old Finns call it "Kuparisaari" — Copper Island. Look to the west. A huge and sweeping panorama of fields and forest — brown and white and green. On the horizon, maybe seven miles away, are the cold deep waters of Mother Superior, the great lake.

Turn to the east now. You are looking at what the old timers call Sunshine Location — the beaver pond and my ranch — and then more forests, with Portage Lake in the distance. On the horizon, ten or twelve miles away, is Mother Superior. Rising above the Mother waters are the strong and sacred Huron Mountains, on the far side of Keweenaw Bay.

This is a place of power and healing. A feast of hundred mile horizons. This is my home, my temple, my family, the mother of my celebration, the magic of my hermitage, my strength, my source, my power, my love. And I sing a song of joyful praise, a prayer of deep thanksgiving. For I am filled with blessings, as rich as rich can be, I am. And I am glad.

Chipmunks & Dragonflies

September 22, 1994. It is the second day of autumn. I am sitting on my deck, drinking coffee and typing. This morning is an ongoing battle between the sun and the fog. So far, the fog is winning. The barn and ancient smokestacks on the horizon appear, disappear, and then emerge again.

The land is a misty watercolor, soft greens, gentle golds, splashed with the vibrant reds and yellows of turning leaves. A mild warm breeze out of the south makes the dewy leaves burble like a shallow stony brook. The breeze rises and falls, not fully awake yet, like me.

But the breeze music is in the background this morning. The main music is the song of robins. There are dozens of them visiting this morning, everywhere I look. In pairs they fly from tree to tree, swooping, diving, banking, and rising. Couples in love, dancing through an autumn fog.

The robins are not here to feed. They are not putting on fat for the long voyage to their winter home. This morning the robins are celebrating. They are celebrating the end of a long and splendid summer. They are celebrating the beauty of the land. They are celebrating the love and friendship that have kept their clan strong and well. Through their song and dance, they are passionately bidding farewell to their summer home. It is a ritual of worship before the long and difficult pilgrimage.

Screech. Scree-ich. Screech. The blue jay clan has arrived, as they do every fall, to feast on Mother Oak's fat acorns. They flutter and fumble on the tiny branches, wrestling with the nuts. When one comes loose, they move to a stronger branch and peck it apart and eat it. As their tummies fill, they begin burying acorns in the yard for later. They take them out into the open lawn, where chipmunks dare not tread.

A pair of chipmunks squirts by, one following the other. They shoot up the hawthorn trunk, jet across the woodpile, and then dive into the tall grass and leaves. They race across the obstacle course fluidly, like a school of minnows, like the robins dancing across the sky. I have no immunity whatsoever to the chipmunk's overwhelming joy. They make me glad to be alive. They force me into outbursts of laughter, and broad warm smiles.

The chipmunks are completely free and alive. They have no wristwatches, no appointments, no debts, and no worries. Their clock is the journey of the Earth around the sun, the passing of the seasons. They know how to live, for they have never forgotten. The chipmunks are my gurus and teachers.

When the chipmunks are done playing, they will return to Mother Oak to gather acorns again. They were filling their cheeks before sunrise, and they will continue until it is too dark to see. The ground is littered with ripped up shells and caps. They race by, heads grotesquely swollen with nuts, like cobras. One has a hole by my porch, and the other has made a home out by the well.

When the chipmunks are tired of playing and gathering, they stop and sit and rest. They sit up and gaze with profound amazement at the perfection of Creation, and offer heart-felt prayers of thanksgiving.

One of the chipmunks has come up on the deck to visit me. He sits, lifts his left front foot, and watches me type. Our eyes meet, and we gaze into one another's souls. His mouth begins moving quickly, but silently. He is talking to me, in a language that my clan has forgotten, but I am somehow able to grasp some of the sentences and phrases.

I can't really translate the words for you, because English is not a language of the spirit. But the chipmunk is sending me poems of love and encouragement and hope. He speaks to me with the simple kind affection that grandfathers use, or tender lovers, or pilgrims overwhelmed with the glory of the divine ecstasy.

All is well, he says. All is beautiful. All is perfect. Lift up your heart in song, for this morning is treasure, a wondrous gift, a rich banquet for the soul. Don't let it sit there and get cold. Dig in, and nourish yourself.

He notices that another chipmunk is now sitting to my right. In the blink of an eye, the chase is on, both giggling as they fly across the grass and up the dead trunk.

The fog has lifted, but the sun is still concealed. Ravens fly by, making the daily progress around their parish, keeping a sharp eye for fresh-killed meat along the road. A single goose flies overhead honking, scanning the farthest horizons for the southbound flight that he has so grievously missed. And now the dragonflies are here too.

The dragonflies are older than the stones. They watched the dinosaurs come and go, and the woolly mammoths. And now they are watching me.

Sacred Migrations

April 18, 1994. It's a warm sunny springtime morning. The temperature is getting close to 50°. Most of the snow is gone. I'm sitting on my front steps, drinking coffee, absorbing the sun's pleasant heat, and giving thanks for the priceless gift of yet another day in this beautiful land. It's nice to be able to sit outdoors comfortably. I feel a bit stiff and sleepy, like a creature who has just emerged from a long winter's hibernation — too much time indoors.

After three days of non-stop high winds, this morning is tremendously calm. The clouds are thin, scattered, high, and slow. To the south, a formation of Canadian geese is flying toward me, honking joyously. From an ancient place deep inside, I smile warmly, with deep satisfaction. The Goose People are returning. Winter lies old, weak, and pale on its deathbed, and spring is big, heavy, and active in the womb, anxious for delivery.

As the flock passes, I notice another flock to the west. Scanning the skies, I see another flock in the distance to the east. The sky is filled with dance and song and hope. The winged nomads are excited to be returning to their northern home. It's time to repair nests, lay eggs, and then nurture them into life. Before long, tiny beaks will burst through delicate shells, and hungry warm fluff balls will celebrate, for the first time, the astonishing perfection of Creation.

The migration of the goose clans continues through my second coffee. And third. And fourth. Wave after wave after wave. Some fly in V's, some in meandering curvy lines, and some in small groups. All of their formations are in a constant state of change and flow. The geese fly high and very fast. All of them will cross Lake Superior today and spend the night deep inside Canada.

The sky music is one of the most ancient songs in the world. It penetrates deep into my heart, mesmerizing and enchanting. I sit transfixed by timeless magic. This land has listened to the songs of the Goose People for the last 10,000 years. As I sit on my front steps, my soul exists in both the present, and the dreamtime of the distant past.

Listening to the music, and the beating flutter of a thousand wings, I close my eyes, and sit in the deep moist fragrant shadow land of a tall radiant dreamtime forest. In my dream, I am dwarfed by a clan of hemlock trees, seven feet in diameter. I am like a tiny red beetle in a huge green cathedral of love.

In the darkness of the forest floor stand large blue-green outcroppings of pure copper, many much taller than me. The crystalline wings, heads, and knobs of the Copper People vibrate from the passing goose music. They resonate, reverberate, and amplify like a hundred harp strings. As the birds fly singing overhead, the metal beings of the land sing back, and loudly. There is nothing but music. The land and sky are passionately embraced in song.

In the dreamtime, the migrations of the Goose People darkened the entire sky, and continued for day after day after day. Their northward journey was a dark and thunderous roar, a winged storm.

Fireflies

June 18, 1995. The morning was slow, dreamy, lost, and confused. I barely slept last night. The heat wave was in its third day now. Comfort was hard to find. Warm coffee, silent mind, gazing away the hours, looking at nothing.

In the afternoon, I hooked up the hoses and watered the garden. Swarms of mosquitoes and gnats took turns feeding on my steaming wet skin. Tiny rainbows sparkled as the arcing shower beat down on the dry soft dust. The garden was almost two weeks old now. Hundreds of tiny seedlings baking quietly in the cloudless blast furnace heat. In a couple moons, these little plants would mature and feed me. I was filled with profound satisfaction. All was well.

By the strawberry beds, a mother painted turtle laid her eggs in a small pit that she had dug. It was a long and careful process. She didn't move as I approached. I sent her my greetings. There have been several turtles laying their eggs here this June. They laid them in sandy places with full southern exposures, where the heat will incubate the unborn. The eggs were laid close to cities of the ant people, so that the hatchlings would quickly find a sumptuous living banquet. When mother turtle was finished laying, she covered the egg pit with bark and grass, making it invisible.

As the sun set, the mosquitoes came out in force. A thousand of them hover in the giant maple tree beside the house. A thousand pairs of beating wings made a clear ringing whine in the air — a loud monotone hum, broken occasionally by menacing pulses. A thousand beings, one mind, all in perfect harmony. The maple was alive with vibrant insect music.

I wandered off to the pond by the old Arcadian mine, and floated in the cool water as the stars emerged from the fading daylight.

Bullfrogs croaked. Warm gentle western breezes called the leaves to dance, called the pond's surface to dance, called the violet pink sunset to dance on the rippling fluid, dancing light hypnotizing my weak tired mind. The water took the heat from my body, and the dirt and sweat and weariness were washed away. I dressed and departed in peace, calm and refreshed.

Home again. I walked bare foot down the road in the soft summer breeze. Shooting stars whizzed across the dark moonless sky. Peepers cheeped and bull frogs croaked and leaves rattled gently on the trees. A small dog barked in the distance.

A million fireflies were out tonight, sitting, floating, drifting, blinking. A million bright green insect lights sparkling below a million twinkling stars. Fond memories from years past comfort and warm me. Firefly memories were always good memories. People who live generous lives, who treat children with exceptional kindness, people who are loved and respected by everyone — these are the ones who return after death as fireflies, to fill the summer nights with beauty and grace.

I walked beyond the woods, stopped, and turned to the right. In the northern sky, an arc of soft green pillars pulsated along the horizon. Aurora borealis. I was overcome with a surge of joy. The night was filled with blessings, and I was the richest person in the world.

Deer & Copper People

Quincy Mine, July 10, 1995. Warm evening, bright sunlight flooding through my window, too intense for reading or writing. It was time to get outdoors and enjoy the setting sun. I got on my bike and headed west, over to the ruins of the Quincy Mine. They've been crushing rock there lately, exposing stones that have been hidden for many decades. Perhaps I would find a treasure.

Hunting for copper was my yoga. By focusing my attention on rocks, my thinker stopped, and I achieved great peace of mind, great stillness and calmness. It always worked. It worked tonight. I became a tranquil animal wandering quietly across the fading evening land.

I spent an hour going through the piles of rocks behind the number four shaft. I found a handful of small copper flakes and nuggets. Nothing special. It was getting dark, so I ripped my weary eyes up off the dusty ancient stones and turned to go back to my bike.

As my eyes swept up the road to the ridge, I saw a deer standing by my bike, twenty feet away. She was a thin doe, probably born last year. She nibbled on some leaves, then stopped to nibble an itch on her shoulder.

It was an odd encounter. Normally deer explode into flight upon sight of a human. This one didn't. I remained calm. I moved my gaze back to the ground, to minimize eye contact. I moved about slowly, like a grazing animal, making no sudden or threatening moves. I've learned that when I stop thinking, the animals are no longer afraid of me. They relax and treat me like a fellow wild one, which is a great honor and blessing.

In a few minutes, two more deer came into view — a second doe, and a young buck with prong horns. Then a second buck appeared. A clan of four.

The first doe was the calmest. She accepted my presence, and went about her way. The other three were anxious and unsure of me. But they seemed to take their cues from the leader, and stayed nervously on the road.

I started to slowly move back up the road, and the lead doe moved with me. I was within twelve feet of her. Two calm creatures sharing a warm and pleasant sunset together. No fear. Magical trust. I was allowed to be their honored guest and friend.

In the distance, I could hear the approach of other humans. They were walking loudly down the rocky road and talking. City humans. They were not calm, quiet, and wild. They had a schedule and a destination. The two bucks and the second doe turned and walked off the road towards the woods.

The invading humans rounded the corner, a young man and woman. They saw the doe, but they kept crashing down the road. As they approached, they saw me standing close to her. At this point, they realized that they had stumbled into the dreamtime. They stopped and gazed upon us, dumbstruck with amazement.

"It sure is tame," the man blurted out. End of magic. The doe bid me farewell and moved off to join her clan. I thanked her and gave her my best wishes.

I exchanged a few sentences with the couple, then departed. This was not a time for words. They were on one channel, and the deer and I were on another. There was no bridge between us, no common language, no possibility of communication.

The deer headed east, the couple went south, and I moved toward the north — glowing, floating, heart bursting with joy and thanksgiving. I had found the treasure, and it was good.

10,000 BC

Silver City, August 24, 1993. From the cliffs overlooking Lake of the Clouds, I walked down to the water level, through a marshland, and up a hill on the other side, primarily through virgin forest. The Porcupine Mountains park consists of 53,000 acres, and most of it is virgin forest. The spirit power of the unspoiled land is incredible. It rushes in to fill the vacuum in my soul, healing and strengthening.

Overhead, a thunderstorm was passing, unseen above the thick forest canopy. Amidst the rumbling, a warm rain showered down through the branches and turned the trail into rivers and lakes. The surface of the steep trail became very slippery.

I took off my shoes and was thrilled by the sensations of wet tree roots, rocks, flowing water, and soft mud. It's incredible how much feeling and intimacy is eliminated by wearing shoes. Shoes disconnect us from the world, and deprive our senses.

I was coming down a steep hillside in a green glade. On one side was a ravine, where a small stream was singing over the rocks. The ground cover and the tree canopy were healthy, living greenery. The rain felt wonderful, and the sound of its trickling down through the branches was cleansing.

I was walking barefoot in the rain in a tremendously sacred and magical place. Ancient images rose up from my unconscious in rushes. I was a man walking through the dawn of time, before farming, before cities, before the dark ages.

The immediate presence of my ancestors was unmistakable. Perhaps this is what Wales was like before the coming of the metal axes, the sheep herds, and the Englishmen. It was a rich experience. As the rain passed, beams of sunlight started shooting through the mist and dripping trees, creating brilliantly-lit islands of green. I wish that words weren't such crude tools.

My babbling inner thinker was quiet for a change, and I was simply a living being walking through a living land. We were all one, beyond time, beyond thought, celebrating the perfection of Creation. The trees and rocks still remembered the passing of Indian hunting parties, the days of clear skies, and a time when the rain was free of acids and metals and herbicides.

With the exception of my clothes and the garbage in my brain, it was 10,000 BC — the place where my heart and spirit feels most at home and alive. It's pretty easy to find 10,000 BC around here. It comes with every fog, it comes every night in the wee hours of the morning when the machines are quiet, it comes when clear waves are crashing on stony unspoiled beaches, it comes with the dancing flames and crackling of a fire late at night.

It comes when picking wild berries in flower-filled meadows under a brilliant blue sky, it comes when the night is filled with bright pulsating shimmering sheets of colored lights, it comes when the music of rain showers fills the air with fresh sweet smells, it comes when a flashing booming thunderstorm is rumbling in the night, and I can't wait for the intense blizzards of the months soon to come.

Contented Gwen

Wales, August 1983. I spent nine days at the National Library of Wales, in Aberystwyth. I studied their genealogical records, in search of my family's roots. I was able to trace my family back to the late 1700s to a hamlet called Cwmbelan.

My research completed, I pedaled away from the coast, into the mountains. Sheep country. Twenty-eight miles into the interior, I entered the village of Llangurig. It was the parish seat of my ancestral home. A tiny village. I walked into the churchyard. Read the headstones. I found it!

The grave of my great, great, great grandfather and my great, great, great, grandmother. "In memory of Edward Rees of Cwmbelan who died April 20, 1849, aged 51 years. Also of his wife, Margaret, who died November 19, 1874, aged 81 years." I stood over them. Home. The church had been the site of family weddings and baptisms. A spirit picnic in the tall grass. Severed connections reattached in the hot sun. I rode two miles down the road to Cwmbelan, adrift in time, absorbing everything. Floating.

Cwmbelan. Only the phone poles, cars, and hardtop on the road were of this century. Otherwise, the land was unspoiled by time. Everything was ancient. A house that my family had lived in was on the left. The wool factory where my great-grandfather had worked as a boy in 1850 was on the right. Ghosts filled the street.

I spent four days in the area. Someone told me to talk to the lady who lived at the old factory in Cwmbelan. I went there and knocked on the big old oak door. From around the corner came an old lady. She was small. Her face was deeply etched with wrinkles. Her glasses were from the 1940s. Around her head, she had a sweatband. She had been gardening. Hello. Hello.

I introduced myself to Gwen and we sat down and talked. There was an aura of wisdom about her. And an aura of joy. She had good eyes — all-knowing, but merry. Her smile was a sunbeam. A beauty.

Gwen lived alone on a pension of just over $20 a week. She did some sewing to supplement her income. Her pedal-powered sewing machine was older than she was. Gwen didn't believe in the modern conveniences. No telephone. No TV. No refrigerator. A friend kept her butter for her.

Everything was just fine in her world. A burbling stream by the front door made lovely water music. The birds chirped. The summer breeze whispered through the trees. Her cats were her good friends. Her garden was her family. The flowers were her children. "Smell this one. Doesn't it have a wonderful perfume?" Yes, magnificent! She smiled. I enjoyed her children, and she liked that.

Gwen was a happy angel. The old factory was her heaven. There was nowhere else she would rather be. Nothing else she would rather be doing. "My mother died when I was twelve. I had to take care of my father for many years. I never had much free time in my life. I was always busy, you see. I'm making up for that now," she smiled. She was one of those rare modern people who knew who she was, and where she was from.

Several years ago, another young American man visited the old factory. He spent a lot of his time traveling and searching. He and Gwen became friends. The American told her that she was the only person he had ever met who was absolutely content with where she was and who she was. He was right. I've met other people who were content, but they have lacked Gwen's radiance. Graceful contentment. A child's sense of wonder. Safe and sound.

She thought about the American's comment. "What is it that people are looking for," she asked. "Is it money? I don't want any money. What good is it? Religion? Most deeply religious people I have met are still afraid of death, so what good is their religion?" Fancy houses? Fancy cars? Fancy things? Meaningless. Her arm swept in a broad arc — the flowers, the stream, the trees, the hillside across the road. "What could be nicer than this? What else do I need?" Nothing, Gwen. Nothing at all. She was radiant.

Always Do That Which Is Best For The Land

Back in 1997, I was active on an online mailing list for the fans of Daniel Quinn's book *Ishmael*. One day, my Irish friend, Robert, posted a question. He wondered how our current value system would function in a turbulent world, during the human downsizing.

His question was this — during the crash, what should you do if you discover a starving man? Should you walk away? Should you feed him? If he survived, and recovered his strength, then what? Would he stay with you until your food was gone? Would he kill you and take your food? What moral obligations do we have to strangers in need?

That night I went out walking in the snowy woods along the ridge of Quincy Hill. An east wind was blowing hard. Gusts pushed on my chest. The dry brown oak leaves were rattling. And as I walked and became more serene, I realized that I was not alone. The spirits of the ancestors were with me. Many of them. Beside me and behind me. There was a feeling of calm and wisdom and peace.

One of them was Gwen. She was wearing a long fur coat and hat — with beautiful embroidery and beadwork. Her long gray hair was blowing in the wind. Her eyes sparkled, even in the darkness, and her smile was a wonderful sunbeam.

Gwen put her hand on my arm, and she whispered some words of ancient wisdom in my ear. And this is what she spoke to me: "These are the words that Robert must hear and remember. This is the most important rule: *Always do that which is best for the land.*"

This is a wonderful rule for living. It's the old thinking, the wisdom of the wee ones, the bright knowledge. If you take care of the land, the land will take care of you. Always do that which is best for the land.

Keeping the Spirit Fire

Walter Bresette passed to the other side on February 21, 1999. Walter lived at the Red Cliff reservation, on the shore of Lake Superior in northern Wisconsin. Vern, Sandy, and I decided to attend the funeral, and we got there the day before. At Walt's house, a spirit fire was burning beside the towering white pine in the front yard. It was lit on the day of his death, and would be kept burning, around the clock, for four days — the day of his burial. The keeper of the spirit fire stood watch until someone was sent to replace him. Someone was always there. Vern and I were asked to keep the fire.

Around the fire were placed four large hunks of wood, arranged in the four directions. On the east and west sides of the fire, blankets were laid on the ground, and on them were laid a square of red cloth and a fist-sized rounded beach stone. People coming to the spirit fire would kneel down, touch the stone, offer a prayer, and get a pinch of tobacco out of the can of sacred herbs. Then, they would move clockwise around the fire to the other side, touch the sacred stone, pray, and toss the offering of sacred tobacco into the sacred flames.

Night arrived, a wet snow started falling, and we talked and listened to the stories of visiting pilgrims. At maybe 10 PM, an Indian man came to relieve us. He told us that Walter was a gift from the Creator — and that the creator was very sparing in sending such loving and giving people to Earth. This man radiated the same sort of strong spiritual power that Walter did, and I deeply admired him.

There were about 500 people at the community center. It was a night of feasting, song, storytelling, and remembering. The next day, Walter was buried with great ceremony. When the process was completed, an eagle flew over the grave, a powerful sign. It was so good to see strong families, and a strong community — one that had carefully preserved and passed along its ancient traditions, customs, and culture. This certainly must resemble the world of my ancestors. Something inside me was awakened.

That funeral was the one and only time in my life that I have experienced being in a functional Earth-centered community, and I will never forget it — my 24 hours in a healthy and sacred normality. I have been longing for that feeling ever since. It felt like home. The

Anishinabe were living in a place where they belonged, and their reverence and respect for that place was loud and clear. They had deep roots there. They knew who they were. What an amazing thing to see.

Ever since that time, I have felt much less comfortable in my own culture, which often feels like an insane asylum. I want to sit by fires on starry nights and listen to the owls talk. Prince Charles said it eloquently: "In so many ways we are what we are surrounded by, in the same way as we are what we eat."

Thunderstorm

July 13, 1995. Sunny warm summer morning, 80° in the kitchen, leafing through a magazine, slowly easing my way into the day. Dense humidity, gray hazy sky, muted sunlight. Silent breeze, trees still, birds singing. Gentle friendly morning of peace.

On my third cup of coffee, I looked up to see a band of fluffy white clouds poking up above the trees, sixty miles away, barely visible through the curtain of hot moist haze. As I sipped my warm brew, slowly, very slowly, the band of clouds climbed higher in the western sky, forty miles away.

Slowly, very slowly, the clouds rose higher and higher. Coffee refill. The sky was in a state of magical transformation. Fluffy, bumpy ice cream clouds had melted in the summer heat, and were now linear, curved, an arc, a colorless rainbow of grays, twenty miles away.

The gray rainbow rose to tower above the land, a monumental skyscraper of moisture, power, and turbulence. Thin wispy lacy clouds led the advance, highlighted by the growing darkness behind them, ten miles away.

Suddenly, the storm's breath arrived, rattling leaves, hissing grass. Then God mashed her foot on the accelerator, and the west wind became a howling gale. The horizon was now pitch black, and the buildings and trees, still lit by the sun, looked eerie and surreal. Five miles away.

The storm then sprang all the way across the sky and devoured the sun. The black sheet of darkness exploded, spewing thousands of huge foaming clouds across the land. The sky was a speeding rush hour freeway of twisting, frothing, raging, hundred ton clouds, all in the eastbound fast lane.

Above the clouds, was the metallic artillery shell rip of lightening bolts slicing through the turbulence and smashing hard into the earth. Boom! R-i-i-i-i-i-i-p BOOM! Large bullets of rain began pelting down. Thunder rumbling, earth shaking, house creaking, windows wet snare drums in the beating rain.

The winds grew to an incredible roar. Trees and bushes were whipping and flailing like flags on rubber poles in a hurricane. Greenery was shaking violently, like the very land was a kettle of vegetation steaming and frothing at an intense boil.

Rain flew sideways in racing, raging, torrential sheets. The road became a riverbed, a million wet bullets exploding on its surface, white-capped waves racing eastward, misting, steaming, splattering, splashing.

Meanwhile, inside my brain, a miracle occurred, a wonderful healing. My civilized thinker was gone with the wind, and I had suddenly become a normal, natural, healthy wild man, a spirit of the wilderness. All of my mental energy was totally focused on the experience of the storm. Thinking was impossible. For twenty or thirty minutes, I was completely out of my mind, and perfectly at one with the world.

Outdoors, the fury gradually subsided, followed by a long, strong, steady shower. Winds slowed, rains diminished, a band of orange light grew in the western sky, and I came to realize that the end of the world was yet to come.

HAPPY ENDING

A Great Healing is Unfolding

I saw the large and powerful thunderstorm coming, when it was still a long ways off. Then it raced in, hit, and hit hard. It brought much needed moisture and restored balance to the hot dry land. The green ones were happy. The birds were singing. Joy was in the air. The storm was good.

In a similar manner, there is now a collapse-driven storm that has filled the sky, large and powerful. In a similar manner this storm will hit, and hit hard. It will be a time of suffering and pain. But when it has passed, the Earth can begin healing. Many Native Americans are glad that the time of healing is near. They have patiently maintained complete confidence that Mother Nature will restore balance. They have never doubted that the disease of industrial society would eventually self-destruct. It was a dead end path.

As the storm proceeds, the industrial way of life will be left rusting peacefully in its grave. Without cheap and abundant energy, it can never be repeated again (at the same scale), thank goodness. Humankind will have nothing but new opportunities — many options, spanning the entire spectrum from suicidal to sustainable.

From the perspective of our wild relatives — the tuna, the wolves, the tigers — the coming human downsizing is not seen as being a disaster. It's the wonderful and thrilling answer to all of their prayers. The crash of the bumbling clumsy civilized humans will be a matter for jubilant celebration and thanksgiving — it's a precious and long-awaited *Great Healing* coming to restore balance and joy to the planet's ecosystems. Hooray! A healthy fever is overpowering a life-threatening infection, and balance will once again be restored.

On the other hand, the people of the consumer worldview will perceive the same healing experience to be a ghastly Armageddon. When human herds experience misery, it's an emergency crisis. But other forms of life barely matter at all. Extinctions do not trigger widespread outpourings of grief, many are not even noticed. This is the mindset that got us into this mess in the first place.

In 1914, the sun never set on the almighty British Empire, and nobody imagined that it would disintegrate within a generation. In 1941, Hitler's Germany appeared to be a rising world-class superpower. In 1971, the collapse of the Soviet Union was on no ones radar, but it fell apart just 20 years later. In 2000, few foresaw that America's days as a superpower were numbered. In 2010, only dreamers imagined that in 500 years, the Corn Belt and the Rust Belt would return to prairie and forest, massive runs of salmon would blast past the ruins of ancient dam sites, vast herds of buffalo would once again thunder across the continent from coast to coast, the Mississippi would be clean enough to drink from, and the American people would become slender, wise, and content — spending their evenings around campfires, singing sweetly under the stars.

Collapse is already in its early stages, and nothing can stop it. It may hit fast (epidemic disease, crop failures, nuclear war, etc.). It may hit slow. Nobody knows. People in rural areas who have a subsistence lifestyle based on muscle power will feel it less. People living in cities and suburbs, who depend on jobs, money, and store-bought food will feel it most.

We can't wish away reality. The avalanche has started, and it will do what it wants to and stop when it's done. So, pay attention, be flexible, and think clearly. We can't make reality disappear by plunging into a sea of distractions. Living in fear is toxic and pointless. The verb *worry* comes from the Old English verb "to strangle." We are alive today, in the current reality. We must accept the cards we were dealt, and play them to the best of our ability.

It's time to enjoy life, to celebrate the magnificence of the gift of existence. At the same time, we should work to use our gifts to contribute to the healing process. The barbaric consumer Dark Age is dying, and the storytellers of the future will enthusiastically describe this as a spectacular victory in the saga of humankind — the last stand of the crazy shoppers. Dawn is breaking out all over. Race outdoors, hug a tree, and enjoy the passing clouds. Life is good! Enjoy it!

The River Has Long Eyes

Lately, I've been living in Oregon. Every day I spend an hour or two by the Willamette River, a sacred relative with a long memory. The river spirits can remember the day when humans first arrived in this region. The Kalapuya people created a way of living that worked, and worked well, for many thousands of years. They were not tillers of the soil. They built no cities. They did not enslave animals. They were not tormented by anxiety and depression. The land, air, and water were clean and healthy. The fish thrived, the forests thrived, the birds thrived, and the animals thrived, including the humans. It was nothing less than heavenly.

The river spirits also remember the day when the white miners arrived. They mined the beavers, they mined the minerals, they mined the forests, they mined the fish, they mined the topsoil, and they created a purely unsustainable way of life that is now moving toward its terminal stage.

It is deeply inspiring to know that this sacred river will also observe the day when the mining era comes to an end, and a long era of great healing begins. The life of the land will recover. Every day, the water spirits remind me that the time of healing is growing near. Better days are coming. Fill your heart with joy and give thanks!

Fairy Music

In the autumn of 2000, I entered into a relationship with a woman in California. In 2003 I sold my Keweenaw ranch. The transition was a big change for me. I was no longer living in the woods, surrounded by wild animals. Now I was living in cities, surrounded by mainstream professional people, in a busy hectic life. I felt like a fish out of water.

Despite the weirdness, returning to the realm of regular people was an important learning experience. I came to see that most folks in this society have never enjoyed an ongoing intimate relationship with their wild relatives, while living in a vibrant wild land — the very essence of a normal, healthy human life. It became vividly apparent that I was not a regular person. I came to see that I was different in a positive way, because I had been very lucky. I realized that my time in the Keweenaw had been a sacred gift.

Most people don't rise every morning in a land of immense beauty and power. Most people do not feel at home in wild places. Most people spend their lives indoors, and have little sense of connection to the natural ecosystem where they live, bouncing like a pinball back and forth between work and home. Many are absolutely terrified of snakes, spiders, wasps, bats, or bears. In essence, many people in this culture live and think like space aliens, spending their lives in an artificial and unhealthy human-built reality.

I also happened to notice that people who had not been intensely studying and discussing the Earth Crisis for 20 years inhabited a mindscape that was totally unrelated to my own. Many seemed to be disconnected — from reality, from life, from the land, from the future, from responsibility. The information that we choose to consume shapes our thinking, and our thinking shapes our behavior.

The finest years of my life, so far, were spent in the Keweenaw. Memories of that experience are burned into my DNA. I will never forget it. My life was so satisfying and rewarding. It seems like the Keweenaw years were the only time in my existence when I was truly and fully alive. Unfortunately, I couldn't afford to keep my house in good repair, and legions of developers were hell-bent on destroying everything I held sacred. Paradise was turning into an ugly subdivision roaring with snowmobiles. It was time to go. So, I moved somewhere far less wholesome. Love has strange power.

The ancient legends of Ireland include stories of people who were carried away to the realm of Fairyland, but had the misfortune of being brought back to the realm of the mortals. Fairy music was said to be unimaginably beautiful. People who heard it were never again the same. They developed a peculiar "spirit-look." They forgot all things, all love and hate, and become distant or melancholy. For the rest of their days, the only sound in their ears was that of the magical fairy harp. If the spell was ever broken, they died.

These stories resonate in my heart. The Keweenaw was my visit to Fairyland, and it was an experience of overwhelming beauty. Then I was brought back to the realm of the normal — an ugly, loud, and crowded world of speeding automobiles, endless entertainment, and furious shopping. This doesn't feel like home. I long for the precious delights of Fairyland, our ancient birthplace and home.

The legends of the fairies remind us of our old ancestral heritage, when people were wild, free, and happy. The world was alive and sacred, and people lived with reverence and respect for the land. But the newcomers — the farmers, loggers, and miners — drove the Good People away. The intruders chopped down their forests, poisoned their wells, exterminated their game animals, destroyed their homes, built filthy cities on their sacred places, and filled the air with soot and noise. The natural world was no longer an enchanted place.

You could imagine that the Earth Crisis is the fairies' revenge. They have summoned all of their magic, and have cast a spell that laid a powerful curse on us. A thousand disasters are circling over our heads. Our silly clever tricks no longer work. We'll pay a dear price for the injuries we've caused to Fairyland. It is right and fair that justice will be done. Hopefully we'll learn and remember and heal. I think we will. A sustainable future with humans in it is not impossible.

Sooner or later, one way or another, the land will heal. The farming, logging, and mining will end, the wee folk will return to their sacred home, and the land will be filled with their music once more.

My life has been a sacred pilgrimage in search of meaning and integrity. It began when I was in high school. I was never able to find integrity in commerce, politics, academia, hedonism, or religion. But I did find it in nature. The living world is sacred and worthy of reverence, respect, and love. It is a solid foundation for an honorable, meaningful, and enjoyable life. Nature is beautiful and good.

During the slave era, many African American songs had themes of escape and deliverance. They expressed a deep hunger for freedom, and a sweet dream of being homeward bound. We are the slaves of consumer society, a violent and soul-killing master. Our chains are psychological.

At core, we long for freedom — a life without clocks or jobs, cars or cities, masters or slaves — a life of love, hope, and celebration. And the rivers dream of freedom, the day when the last dam falls apart. And the forests dream of freedom, the day when the cutting stops. Everything everywhere wants to be free, and freedom day is coming. The cruel old master is sick and feeble, and his days are growing short. In the other world, the spirits of our wild ancestors are filled with joy. The Earth shall be free once more, and forever.

The Sustainability Paradox

Sustainability is the ideal and the norm. For most of human history, we lived in a relatively sustainable manner. But we fell sharply out of balance when we began domesticating plants and animals, and our ecological trajectory has been downhill ever since. With each passing century, we have been moving farther and farther from our sustainable origins. Our progress, our science, and our technology are speeding our decline, our race to destruction. Yet we celebrate progress as being beneficial. This represents a huge mental disconnect — and a huge opportunity for growing and healing.

The Native Americans found the European settlers to be very scary people, because the whites were unbelievably out of balance. There was nothing that the colonists hesitated to destroy — forests, fisheries, societies. Their craziness was so intense that it was contagious. Before long, the Indians were exterminating the beavers in order to trade for guns, whiskey, and other odd things (understanding that if they didn't do it, the whites would).

Indian visionaries like Neolin and Tenskwatawa called out bold warnings to their people. The Europeans were the evil children of the Great Serpent. The Indians could only save themselves by rejecting the Christian way of life, and returning to their traditional ways. They needed to rid themselves of domesticated animals, woven clothes, alcohol, metal tools, and selfish materialism. They needed to return to animal skin clothing, the bow and arrow, and subsistence hunting.

These instructions were remarkably intelligent, but they only went half way, because the settlers had zero interest in returning to the tribal traditions of their wild and free ancestors. Intelligent and crazy do not mix. The unfortunate reality is that lads with guns will usually subdue lads with bows and arrows. Lads with plows and cows will beat the deer hunters. Lads with horses will beat the foot soldiers. Lads with railroads, rockets, jet fighters, and battleships can pretty much do whatever they please — in the short term. The root of the problem is the crazy worldview that motivates people to conquer, exploit, and oppress, by any means possible. A crazy society is no fun for anyone.

A dozen years ago, an internet philosopher named Joe Gaglio kept asking a disturbing question about the industrial society that is destroying the ecosystem: "How can we expect to stop them by emulating those that have been destroyed?" But those who have been destroyed include those who lived sustainably. What could be more intelligent than emulating sustainable ways of living? How could you effectively eliminate or illuminate the destroyers?

They will continue to destroy throughout the collapse, and keep living destructively until: (a) there's nothing left –or– (b) they wake up. The only effective long-term solution is to outgrow our crazy worldview. If we can remember how to be human beings again, our worldview will heal.

I once read a book by the German historian Adam of Bremen, who wrote almost a thousand years ago, in the 1070s. He was a literate and well-educated cleric, in a world with very few books, long before the printing press. I probably read more in a month than he read in his life. In Adam's descriptions of northern Europe, he described, in a matter-of-fact manner, the Amazons who lived in the land of women, the Troglodytes who could run faster than wild beasts, and the one-eyed Cyclops.

Today's educated readers snicker at these silly old superstitions. We have replaced them with an equally silly new superstition: the Technology Fairy, who will magically provide us with a totally sustainable future that includes automobiles, computers, cell phones — requiring absolutely no sacrifices in our way of living — and endless economic growth, too!

At the same time, more intelligent trends are also emerging. In consumer society, most of us are now literate, and we have access to unimaginable quantities of information. We are, by far, the best informed generation in all of human history. In the last 30 years, huge and positive gains have been made in the way we perceive issues of ecology. In my wildest fantasies, I wish that these gains in consciousness would be happening much, much faster, but every gain helps. Every gain is good.

We are not on a dead end path because of a tragic shift in our genetic evolution. We are not the helpless victims of irreparable chromosome damage. In fact, we are on a dead end path because we have been taught dead end thinking. Errors of thought are absolutely reparable, but big transitions can take generations.

We can deliberately move straight toward a sustainable future, or we can cling to our toxic addictions and flounder away many years foolishly chasing the false promises of ersatz sustainability. Either way, the Earth will return to balance, with or without humankind. The good news is that all paths will eventually lead to a happy ending to the horror story of human-generated ecological destruction. The madness cannot survive indefinitely. Nothing unsustainable can survive indefinitely. We are not on a wise path. My advice would be to turn around, and begin our long walk home, to our ancestors' fires.

The powerful healing magic of the Ghost Dances will eventually overcome the powerful destructive curse of human foolishness, one way or another. The filthy carpet of civilization will be rolled away. Fairyland will rise and shine once more. The wee folk will return to their ancient homes, and fill the world with joyful music — love, hope, and celebration.

Coyote Lessons

One night, long ago, I was out walking in the woods near my Keweenaw ranch, a cool night with lots of stars. Suddenly a choir of coyotes exploded into a glorious wild serenade. Loud passionate cacophony. Yip-yip-yip! Yippity-yip! Yooowl! Hooowl! Oooooo... Wheee!!! High voltage six-part harmony. Coyotes sing with astounding enthusiasm! They celebrate the magnificence of their existence with every ounce of their being. Coyotes do not understand the meaning of subtlety or reserve. They shout and scream and yelp and howl — at the very tops of their lungs — about nothing but freedom, wildness, joy, and the beauty of the Earth.

Coyotes thrill me! It is redundant to speak the words *wild coyote*, because all coyotes are wild. All coyotes are free. Coyotes have never given away their power to the machine that is destroying the world. Coyotes have never allowed themselves to be tamed, collared, penned, owned, controlled, bred, neutered, numbered, tagged, weighed, vaccinated, or slaughtered. Coyotes have never forgotten who they are, or their sacred place in this sacred world.

A coyote doesn't come when you call him, sit when you tell him, or stand up and beg for a treat. A coyote doesn't race in from the woods when the school bell rings — or the church bell, or the factory whistle. A coyote is not owned and controlled by human masters. He has no interest in becoming our friend. He is living as coyotes were intended to live — in their ancient and sacred manner, with great pride, dignity, and integrity. He is strong, whole, complete, at one.

Coyotes have so much to teach us. We have so much to learn. May the ancestors protect you and guide you.

SELECTED READINGS

Anderson, M. Kat, *Tending the Wild — Native American Knowledge and the Management of California's Natural Resources* (2005) describes how Indians manipulated their ecosystem to increase the production of the things they needed.

Arseniev, Vladimir K., *Dersu the Trapper* (1941) describes the life of Dersu, one of the last wild Siberian hunters, and his amazing understanding of the living ecosystem.

Ashworth, William, *The Late Great Lakes* (1986) does a powerful job of describing the ecological destruction of the Great Lakes. Great book!

Axtell, James, *The European and the Indian* (1981) describes the interactions between Native Americans and the European colonists. Many settlers abandoned white society to enjoy a happier life among the indigenous people.

Basalla, George, *The Evolution of Technology* (1988) discusses why some technology goes extinct, and other technology is developed further. Sacred myths are challenged.

Blythe, Ronald, *Akenfield — Portrait of an English Village* (1969) is an oral history of the old-timers living in a rural English farming village.

Bresette, Walter and Whaley, Rick, *Walleye Warriors* (1994) describes the struggle to preserve fishing rights in northern Wisconsin in the 1980s. Additional material on Walter can be found at www.protecttheearth.org.

Bright, Michael, *Man-Eaters* (2000) provides hundreds of brief stories about man-eating tigers, lions, sharks, snakes, crocodiles, and many others.

Brown, Lester R., *Plan B 3.0 — Mobilizing to Save Civilization* (2008) provides a stunning description of our biggest problems, and a plan for rapidly solving them. The book can be downloaded free from www.earthpolicy.org.

Brown, Lester R. and Wolf, Edward C., *Soil Erosion: Quiet Crisis in the World Economy* (1984) carefully describes one of the biggest problems that few are talking about — soil destruction.

Brown, Lester R., *Who Will Feed China?* (1995) predicts that Chinese population growth will lead to ever-increasing food imports, eventually creating a demand that the food exporters can no longer meet.

Brown, Tom Jr., *The Tracker* (1978) describes the rich and intimate relationship with the family of life that the author learned from a traditional Native teacher, Stalking Wolf. Great book! If you enjoy this one, read his many other fascinating books.

Campbell, Colin J., *The Coming Oil Crisis* (1997) discusses the history of petroleum discovery and production, and predicts that production will peak by 2010 or so. After the peak, production will decline gradually, year after year.

Carter, Vernon Gill and Dale, Tom, *Topsoil and Civilization* (1974) provides a history of the rise of civilization, and the destruction of topsoil, covering 8,000 years or so. Great pictures! Excellent book! Full text is available for free on the Internet.

Clarke, W. C. and Thaman, R. R., eds., *Agroforestry in the Pacific Islands: Systems for Sustainability* (1993) provides an island-by-island examination of agroforestry, and recommendations for development planners.

Clover, Charles, *The End of the Line* (2006) is an excellent book on industrial fish mining, and the destruction of oceanic fisheries.

Cohen, Mark Nathan, *Health and the Rise of Civilization* (1989) describes how the development of civilization diminished human health.

Collier, Richard, *The Plague of the Spanish Lady* (1974) describes the horrors of the global influenza pandemic of 1918.

Coon, Carleton S., *The Hunting Peoples* (1971) provides a nice collection of information on the nomadic forager lifestyle.

Cox, Stan, *Sick Planet — Corporate Food and Medicine* (2008) delivers a blistering critique of organic agriculture, and a provocative formula for weaning the US from fertilizer use.

Cronon, William, *Changes in the Land* (1989, originally published in 1983) describes how New England changed ecologically following the arrival of European settlers.

Crosby, Alfred W., *Germs, Seeds, and Animals: Studies in Ecological History* (1994) discusses the ecological impacts of the emigration of Europeans during the colonial era.

Crosby, Alfred W., *Throwing Fire — Projectile Technology Through History* (2002) discusses how the evolution of weaponry has shaped human history, and now threatens to end it.

Dasmann, Raymond F., *The Destruction of California* (1965) describes the growing ecological decimation of California following the arrival of civilized people, and questions how complete destruction can be avoided.

Davis, Mike, *Late Victorian Holocausts* (2001) describes the horrors of overpopulation and imperialism during 19th century El Niño droughts in China, India, and Brazil. This is a book that you will never forget.

Deloria, Vine, *Red Earth, White Lies* (1995) discusses a number of issues where science is at odds with Native American traditions, and questions the integrity of science.

Devall, Bill, editor, *Clearcut: The Tragedy of Industrial Forestry* (1993) examines the ecological horrors of the modern logging industry. Stunning photos!

Diamond, Jared, *Collapse* (2005) describes the mental and physical processes of societal self-destruction in old cultures and contemporary ones.

Diamond, Jared, *Guns, Germs, and Steel* (1997) describes the spread of civilization, how and why.

Diamond, Jared, *The Third Chimpanzee — The Evolution and Future of the Human Animal* (1992) explores how we became what we are today. We experienced a Great Leap Forward, which also provided us with the tools for self-destruction.

Diamond, Stanley, *In Search of the Primitive — A Critique of Civilization* (1974) analyzes modern society, explains why mental illness is so common, and provides a nice discussion on the myth of progress.

Ehrenreich, Barbara, *Bright-Sided — How the Relentless Promotion of Positive Thinking has Undermined America* (2009) describes how the cult of positive thinking has dangerously altered our mental processing.

Ehrlich, Paul and Ehrlich, Anne, *Earth* (1987) is an excellent primer on ecological history. *The Dominant Animal - Human Evolution and the Environment* (2008) expanded and updated the discussion.

Fagan, Brian, *The Little Ice Age — How Climate Made History 1300-1850* (2000) describes how climate shifts affected European agriculture and society.

Farb, Peter, *Man's Rise to Civilization* (1968) describes the negative effects of civilization upon various indigenous groups in North America.

Flannery, Tim, *The Eternal Frontier* (2001) provides a thorough ecological history of North America. It argues that the megafauna extinctions were primarily the result of human hunting.

Flannery, Tim, *The Future Eaters* (1995) discusses the ecological history of Australia and surrounding regions. The early nomadic foragers had an immense ecological effect on the land, plants, and animals. It wasn't until after the fat of the land had been destroyed, and survival became more challenging, that humans developed a culture of reverence for the land.

Fleay, Brian, *Decline of the Age of Oil* (1995) examines the future of oil. This book provides a great analysis of the relationship between petroleum and the human food supply.

Forbes, Jack D., *Columbus and Other Cannibals* (1992) talks about civilization as a contagious and destructive spiritual disease.

Fraser, Evan D. G. and Rimas, Andrew, *Empires of Food* (2010) discusses the role of agriculture in the rise and fall of civilizations, and our current precarious state.

Freinkel, Susan, *American Chestnut — The Life, Death, and Rebirth of a Perfect Tree* (2007) describes the blight that ravaged the American chestnut, and the efforts to save it.

Freuchen, Peter, *Book of the Eskimos* (1961) provides a touching portrait of the traditional Eskimo way of life. Great book!

Garrett, Laurie, *The Coming Plague* (1994) discusses current trends in contagious diseases, and suggests what the future may hold for us.

Glendinning, Chellis, *My Name Is Chellis & I'm in Recovery from Western Civilization* (1994) provides strategies for recovering from the emotional traumas of being raised in industrial civilization.

Gore, Al, *Earth in the Balance* (1992) is a primer on the Earth Crisis. It was followed by *Our Choice — A Plan to Solve the Climate Crisis* (2009) which focuses on climate change and how to fix it.

Gottfried, Robert S., *The Black Death* (1983) describes the social aspects of the era of bubonic plague.

Gray, John, *Straw Dogs — Thoughts on Humans and Other Animals* (2002) helps us understand the workings of the civilized human mind, in a powerful philosophical critique of progress and humanism.

Greer, John Michael, *The Long Descent* (2008) provides insights on the currently unfolding collapse of global industrial civilization.

Greger, Michael, M.D., *Bird Flu — A Virus of Our Own Hatching* (2006) describes the current threat of pandemic influenza, and the health risks associated with raising domesticated livestock, waterfowl, and poultry.

Gunther, John, *Inside Africa* (1953) discusses the history and current affairs of every country in Africa, including the friction between black and white society.

Hanson, Jay, www.dieoff.org provides a huge online collection of essays and reports, from reputable scholarly sources, that document why we are heading for big changes in the coming years.

Heinberg, Richard, *Peak Everything — Waking Up to the Century of Declines* (2007) describes a future of decline, because a number of finite resources are coming into short supply.

Heinberg, Richard, *The Party's Over* (2003) discusses the fossil fuel era, and the consequences of the end of the era of cheap energy.

Herzog, Hal, *Some We Love, Some We Hate, Some We Eat* (2010) discusses the quirky relationship that humans have with other animals.

Hornaday, William T., *Our Vanishing Wild Life* (1913) describes the horrifying destruction of wildlife in North America. It provides a glimpse of how abundant wildlife used to be in this nation. The complete text of this book is available free online. *The Extermination of the American Bison* (1889) describes the process of slaughtering the vast herds of buffalo. The complete text of these books is available free online.

Jackson, Wes, *New Roots for Agriculture* (1985) is a critique of agriculture and the society that supports it. His research on perennial agriculture is discussed. He expanded on this theme in *Altars of Unhewn Stone* (1987), and *Consulting the Genius of the Place* (2010).

Jenkins, Joseph, *The Humanure Handbook* (2005) provides a thorough discussion on composting excrement and using it for fertilizer.

Jensen, Derrick, *A Language Older Than Words* (2000) examines the reasons behind the Earth Crisis, especially the violence, and contemplates how to respond to it in an intelligent manner. This was followed by *The Culture of Make Believe* (2002), *Strangely Like War* (2003), *Endgame* (2006), *Thought to Exist in the Wild* (2007), and others. Jensen is a warrior on the radical fringe.

Johnson, Steven, *The Ghost Map* (2006) describes mid-19th century cholera epidemics in London, and how the cause of cholera was discovered.

Josephy, Alvin M., Jr., *500 Nations — An Illustrated History of North American Indians* (1994) describes the conquest of North America, from Columbus to Custer. *The Patriot Chiefs* (1961) focuses on the heroic war chiefs who fought to bring down civilization and preserve the American way of life.

Jung, Carl Gustav (edited by Violet Staub de Laszlo), *The Basic Writings of C. G. Jung* (1959) provides important passages from Jung's many published works.

Jung, Carl Gustav (edited by Meredith Sabini), *The Earth Has a Soul* (2008) provides a collection of excerpts from Jung's writings on the subject of nature, and humankind's disconnection from nature.

Junge, Traudl, *Until the Final Hour — Hitler's Last Secretary* (2002) provides an inner circle perspective of Hitler's final years — life in the Nazi worldview.

Keith, Lierre, *The Vegetarian Myth* (2009) provides a critique of vegetarianism, agriculture, civilization, patriarchy, low-fat diets, and soy foods. Her mental and physical health was injured by eating a vegan diet for 20 years.

King, Franklin Hiram, *Farmers of Forty Centuries — Or Permanent Agriculture in China, Korea, and Japan* (1927) describes the low-tech, labor-intense subsistence agriculture practiced in eastern Asia in 1909. This book can be downloaded from Project Gutenberg.

Kirch, Patrick Vinton and Yen, D. E., *Tikopia — The Prehistory and Ecology of a Polynesian Outlier* (1982) describes Tikopian culture and history.

Krech, Shepard, *The Ecological Indian* (1999) discusses the ecological impacts of Native American societies, challenging the tradition that they were eco-saints.

Lame Deer, John (Fire) and Erdoes, Richard, *Lame Deer — Seeker of Visions* (1972) describes Lakota culture and history, including the Ghost Dance movement.

Liedloff, Jean, *The Continuum Concept — In Search of Happiness Lost* (1977) describes the social and psychological differences between civilized US families and the indigenous families in South America. This is a great book.

Lillard, Richard G., *The Great Forest* (1947) describes the history of the destruction of North America's forests.

Livingston, John A., *The John A. Livingston Reader* (2007) contains two essays: The Fallacy of Wildlife Conservation (1981), and One Cosmic Instant: A Natural History of Human Arrogance (1973). He presents an important discussion on the emergence of human superiority, and the terrible toll that humanism is taking on the planet's ecosystems.

Livingston, John A., *Rogue Primate — An Exploration of Human Domestication* (1994) appends an additional ten years of thinking to his earlier theories.

Lowdermilk, Wayne C., *Conquest of the Land Through Seven Thousand Years* (1948) is an excellent overview of the history of soil destruction, and how to prevent it. This booklet was published by the USDA, and is available on the internet, with stunning photos.

Levy, Stuart B., M.D., *The Antibiotic Paradox* (2002) describes the history of antibiotics and the growing problems caused by their rampant misuse.

Mann, Charles, *1491 — New Revelations of the Americas Before Columbus* (2005) provides a stimulating and up-to-date overview of what the Americas were like just prior to Columbus.

Manning, Richard, *Against the Grain — How Agriculture has Hijacked Civilization* (2004) discusses the history of agriculture, with emphasis on modern industrial agriculture.

Manning, Richard, *Grassland* (1995) provides a wealth of information on prairie ecosystems. This was followed by *Rewilding the West — Restoration in a Prairie Landscape* (2009) explores the possibility of restoring wild animals, including buffalo, to the grasslands of the West, living in a fence-free open range.

Manning, Richard, *One Round River* (1997) describes the pointless environmental destruction caused by gold mining and open pit mining.

Margolin, Malcolm, *The Ohlone Way* (1978) describes the Ohlone people of San Francisco Bay, who created a fairly peaceful complex society without agriculture. He also wrote *The Way We Lived* (1993), a collection of oral history on California Indians.

Marsh, George Perkins, *Man and Nature* (1864) is the book that started the ecology movement. Marsh recognized that ancient civilizations all self-destructed. This is an important book that has tragically been ignored by the high priests of education for a century and a half. The complete text of this book is available free online.

McNeill, J. R., *Something New Under the Sun — An Environmental History of the Twentieth-Century World* (2000) discusses 20th century problems, and their historic roots.

Meadows, Donella H., *Limits to Growth* (1977) provides genuine scientific evidence that the world and resources are finite. It was a breakthrough book for its time.

Mintz, Sidney W., *Sweetness and Power — The Place of Sugar in Modern History* (1985) provides a history of sugar production, from a British perspective.

Molyneaux, Paul, *Swimming in Circles* (2007) provides an insider's perspective on industrial aquaculture, and its many serious problems.

Montgomery, David R., *Dirt — The Erosion of Civilizations* (2007) provides an up-to-date history of civilization's war upon the soil.

Montgomery, Sy, *Spell of the Tiger* (1995) describes how tigers protect the ecosystem of the Sunderbans region of India, and why the natives worship them.

Mowat, Farley, *People of the Deer* (1951) discusses the hunting people of northern Canada, and how their ancient way of life was destroyed by the arrival of the fur traders and civilization. He also wrote *The Desperate People* (1959) and *Never Cry Wolf* (1963), which explore this story further.

Mumford, Lewis, *The Myth of the Machine* (1966) is a brilliant and complex commentary on the origination, operation, and faults of what he contemptuously calls the megamachine — industrial civilization. Two volumes. Not easy to read.

National Research Council, *Microlivestock: Little-Known Small Animals With A Promising Economic Future* (1991) discusses birds, rodents, and lizards that can be raised on a small scale for food.

Nerburn, Kent, *Neither Wolf Nor Dog* (2002) describes how "Dan," a traditional Lakota elder, perceives the white civilization that has invaded his homeland — with gloves-off honesty. A powerful book. The follow-up book is *The Wolf at Twilight* (2009) which describes the cultural genocide performed at Indian boarding schools.

Nohl, Johannes, *The Black Death* (1926) describes the social aspects of the era of bubonic plague.

Orlov, Dmitry, *Reinventing Collapse — The Soviet Example and American Prospects* (2008) describes the social effects of the economic collapse of the USSR, and how Americans can prepare for their collapse. A provocative book.

Perlin, John, *A Forest Journey* (1991) discusses the history of forest destruction, and its consequences. This is a must-read book for folks who believe that the solution to the Earth Crisis is to turn the clock back 500 years.

Petersen, David, *Heartsblood — Hunting, Spirituality, and Wildness in America* (2000) compares the beauty of ancient sacred hunting with modern high-tech body count hunting.

Pollan, Michael, *The Omnivore's Dilemma* (2006) provides a fascinating analysis about the way we produce and eat food — with special emphasis on industrial-scale agriculture, and the harms it causes. Great book!

Pollan, Michael, *In Defense of Food* (2008) recommends how we should eat to recover our health. This means avoiding the standard Western diet, which is killing us.

Ponting, Clive, *A New Green History of the World* (2007) is an encyclopedic ecological history of the world, jam packed with facts, a treasure chest of information. Great book! It updates *A Green History of the World* (1991).

Postel, Sandra, *Pillar of Sand* (1999) discusses the growing ecological harms caused by the use of irrigation in agriculture, and predicts serious problems ahead.

Price, Weston A., *Nutrition and Physical Degeneration* (1939) presents dramatic evidence that people who eat traditional diets are far healthier than those who eat modern Western diets that major in processed foods — many unforgettable photos. Full text is available for free on the Internet.

Prince of Wales, Charles, *Harmony — A New Way of Looking At Our World* (2010) presents a stimulating discussion on the defective modernist worldview, and contemplates a healthier future that is in harmony with nature.

Pritchard, Evan T., *No Word for Time* (1997) discusses traditional Micmac wisdom.

Quinn, Daniel, *Ishmael* (1992) is a rich groundbreaking novel describing the truth about the cultures of civilization, and how they are destroying the world. Great book! He further developed this discussion in *The Story of B* (1997) and other books.

Raphael, Ray, *Tree Talk* (1981) and *More Tree Talk* (1994) provide an excellent introduction to the methods and problems of industrial forestry. Suggestions for kinder, gentler forest destruction are offered.

Reader, John, *Man on Earth* (1990) describes the lifestyles of various types of premodern cultures.

Reader, John, *Potato — A History of the Propitious Esculent* (2009) describes the history of the potato, and how this plant has changed human societies and world history.

Reisner, Marc, *Cadillac Desert* (1987) describes the history of dam building in America, in a lively and enjoyable manner. Fascinating dust storm section.

Rixson, Denis, *The Small Isles — Canna, Rum, Eigg, and Muck* (2001) describes the enclosure of the Small Isles and their transition into the modern world.

Roberts, Paul, *The End of Food* (2008) examines our food system, and raises serious doubts about our ability to feed ten billion in a few decades.

Rowe, Stan, *Home Place — Essays on Ecology* (1990) and *Earth Alive — Essays on Ecology* (2006) describe how our culture has turned us into ecological destroyers.

Ruiz, Don Miguel, *The Four Agreements* (1997) discusses how to throw off the chains of domestication and become a happy, clear-thinking, wild and free person.

Ruppert, Michael C., *Confronting Collapse* (2009) discusses the geopolitics of Peak Oil, and the reality that we are beyond the point of no return — industrial civilization is in the process of collapsing.

Ryan, Christopher and Jethá, Cacilda, *Sex at Dawn — The Prehistoric Origins of Modern Sexuality* (2010) describes how our adaptation of agriculture radically altered human sexuality. Women were downgraded from equals to property.

Shepard, Paul, *Coming Home to the Pleistocene* (1998) describes how humankind is genetically designed to thrive in wild lands, living in small bands, eating wild food. Modern living destroys us physically, mentally, and spiritually. Shepard carefully examines the horrors associated with the domestication of plants and animals.

Shepard, Paul, *Nature and Madness* (1982) provides a stunning and original psychological and spiritual history of humankind, as we shifted from foraging to gardening to civilization. Important book, but hard to read.

Shepard, Paul, *The Others — How Animals Made Us Human* (1996) describes the ancient sacred relationship between humans and other animals, and the tragedies of animal domestication. Great book!

Shepard, Paul, *The Tender Carnivore* (1973) describes human origins, and the problems of civilization. Shepard envisions a future where we blend the primitive way of life with high technology — a notion he later dropped, as his understanding grew.

Smith, Joseph Russell, *Tree Crops — A Permanent Agriculture* (1929/1987) discusses many types of trees that can be used to provide food for humans and livestock.

Speer, Albert, *Inside the Third Reich* (1970) describes life in the inner circle of Hitler's government — a very civilized and industrial regime. In prison, Speer's conscience deepened, as he wrote this touching story.

Spellberg, Brad, *Rising Plague* (2009) discusses antibiotics, the growing number of drug-resistant microbes, and the bleak outlook for antibiotic research.

Stannard, David E., *American Holocaust* (1992) provides a detailed and stunning description of the extermination of up to 100 million Native Americans.

Stanton, William, *The Rapid Growth of Human Populations 1750-2000* (2003) provides an outstanding discourse on population growth, from a perspective refreshingly free of political correctness.

Trudell, John is a poet, performer, activist. He's an articulate, brilliant, creative thinker, but he's not a book writer. There are written and video interviews on the internet. There is also a documentary movie about his life titled *Trudell*.

Turnbull, Colin M., *The Forest People* (1961) is a lovely story of the culture and folkways of the Pygmies. It shows us a culture vastly superior to our own, ecologically and socially — and fun and beautiful, too. Great book!

Turnbull, Colin M., *The Human Cycle* (1983) provides a lovely comparison of the cultures of India, the Pygmies, and upper class England. Great book!

Turnbull, Colin M., *The Mountain People* (1972) describes the tragedy of the Ik people, whose ancient hunting lands were confiscated to make a wildlife preserve. This is a dark and haunting story, and an all-time favorite of hard-core misanthropes.

van der Post, Laurens, *The Heart of the Hunter* (1961) provides a fond description of the Bushmen of South Africa. He sees their culture as being far superior to civilized culture.

Ventura, Michael and Hillman, James, *We've Had a Hundred Years of Psychotherapy — and the World's Getting Worse* (1992) discusses the psychological aspects of healing the Earth Crisis.

Weatherwax, Paul, *Indian Corn in Old America* (1954) is a history of the corn culture of indigenous Americans, including many illustrations.

Weisman, Alan, *The World Without Us* (2007) is a delightful and informative book that discusses how the world would change if all humans suddenly disappeared.

White, Lynn Jr., *Medieval Technology & Social Change* (1962) describes the technological advances of the Middle Ages.

White, Richard, *The Roots of Dependency* (1983) describes the methods by which three Native American tribes were forced into dependency on white society.

Winslow, Charles Edward Amory, *Man and Epidemics* (1952) provides a thorough discussion on the relationship between civilization and the spread of deadly contagious diseases.

Wolff, Robert, *Original Wisdom* (2001) describes primitive healing, and life among the Sng'oi, who taught him the art of using his ancient intuitive mind. Great book! His website is www.wildwolff.com.

Woodham-Smith, Cecil, *The Great Hunger — Ireland 1845-1849* (1962) describes the Irish potato famine, a horror story generated by overpopulation and monoculture farming, and made worse by British cruelty.

Youngquist, Walter, *GeoDestinies* (1997) discusses current trends in the consumption of geological resources (especially petroleum), the impacts of an exploding population, and the approaching end of the era of cheap energy. An excellent book.

Zinn, Howard, *A People's History of the United States* (1980) describes the behavior of Europeans in the New World. They were not nice to the indigenous folks, or to the lower classes of all races. Great book!

INDEX

A

Adam of Bremen, 35, 94, 309
Adrian, author's middle name, 97
Aesir (Norse gods), 260
Agriculture
  Agriculture not necessary, 218
  Cuba struggles with organic, 108
  Diseases of farming, 142, 179
  Ecosystem destruction, 194
  Feeding ten billion, 154, 162
  Harshness of Medieval farming, 175
  Last 200 years, 102
  No-till farming, 146, 148
  Oil-powered agriculture, 108
  Organic farming harms, 78
  Origins, 76, 118
  Phase out agriculture in future, 110, 216
  Return to muscle-powered farming, 108
  Risks, 79, 151
  Slash & burn, 171, 172, 183, 184
  Source of conflict and war, 65, 81
  Sustainable agriculture?, 165
Agroforestry, 207, 214, 313
Akwesasne, 111
Alanis, A. J., 232
Amish, soil mining, 173
Ancestors
  Author's ancestors, 24–44
  Importance of good ancestors, 23
  Reconnection with, 43, 272
  Sustainable in past, 42
Anderson, M. Kat, 206
Anishinabe tribe, 14–22, 14, 21, 22, 41, 43, 85, 277
Annual plants, 117
Antibiotics, 73, 104, 189, 190, 196, 230
Aquaculture, 188–93
Aral Sea destroyed, 123
Arseniev, Vladimir, 225
Arsonist culture, 253
Axtell, James, 87

B

Baby farm, infanticide, 242
Bageant, Joe, 270
Ballistic weapons, 54
Bannock tribe, 74
Basalla, George, 51, 216
Beaver Wars, 63
Big brains, emergence of, 53
Big heads disturb harmony, 50
Billie, Danny, 264
Birth customs, 282
Black Death. *See* Bubonic plague
Blind optimism, 276
Blue baby syndrome, 139, 199
Blythe, Ronald, 176
Borlaug, Norman, 130, 149
Bow and arrow, 18, 43, 54, 58, 187
Brainstorms, dangerous, 59
Breeding plants, 103, 152, 153, 155
Bresette, Walter, 14, 20, 43, 251, 277, 281, 298
Bright, Michael, 224
Brown, Tom, 4
Bubonic plague, 38, 40, 42, 213, 228, 229, 230
Buffalo
  Hunting, 74, 182, 194, 195, 257
  Original range, 195, 210
  Restoration of wild herds, 199, 263, 304
  Wisdom, power, sense of humor, 67
  Wisent, 26, 34, 73
Burial customs, 282
Burning by Native Americans, 47, 194, 210

C

CAFO. *See* Meat
Cancer, 123, 139, 158, 161, 197, 274
Cats, domesticated, 68
Cave paintings, 51
Cherokee tribe, 88
Cherokees, eastern US tribe, 65
Chestnut die-off, 211
Chickens, 73, 180, 202, 228
Cholera, 32, 36, 42, 136, 142, 213, 230
Civilization
  Arrival in Europe, 22
  Arrival in Keweenaw, 21
  Birth of civilization, 84
  Devastation of forests, 95

Failure of Native American civs, 18
Hated by God, 86
Industrial civilization, 96
Major errors, 85
Clear thinking, 279
Climate change, 36, 106, 124, 162, 178, 197
Clovis points, 56
Coal, 96, 105, 259
Collapse
Benefits of, 264, 303
Causes of, 262
End of cheap energy, 105
Fear of, 303
Unavoidable, 113, 304
Collective cleverness, 60
Commons, 30
Community restored, 265
Complex language, 53
Conscious mind, 247, 250
Consumers, 101, 248, 267
Copper Man, 20
Copper mining, 21
Copper people, 16, 284
Corn
Corn blight, 80
Fertilizer benefits, 130
Health effects, 187
Heavy feeder, 184
Population & conflict, 186
Source of conflict, 65
Spread of corn farming, 181
Three sisters farming, 181
Cover crop, 129
Cox, Stan, 132, 148, 175
Coyotes, 69, 70, 311
Crazy Horse, 258
Cronon, William (three sisters), 184
Crop rotation, 36, 129
Crosby, Alfred, 104
Culture
Cultural evolution, 52
Cultural information, 52
Cwmbelan, Wales, 26–28, 32, 99, 296

D

Davis, Mike, 80
de Condorcet, Marquis, 61, 236
Dead zone, 137–39, 138, 139
Defensive pessimism, 276
Deforestation, 35, 36
Deloria, Vine, 56

Denial, 249, 254, 255
Depression, rare in primitives, 251
Dersu the Trapper, 225
Deserted medieval villages (DMV), 31
Diabetes, 161
Diamond, Jared, 50, 171, 216, 253
Die-off
Benefits of die-off, 262
Die-off of reindeer, 223
Directed thinking. *See* Conscious mind
Disease, contagious, 227
Disenchantment of nature, 46, 307
Dogs, 67, 68, 69, 141, 228, 267
Domestication
Animals, 66–75
Birth of contagious disease, 227
Grains, 77
Metals and minerals, 88, 89, 90
Of humans, 75, 84
Ducks, 180, 202, 227
Dust Bowl, 59, 145

E

Earthworms, 203
Easter Island
Slash and burn deforestation, 173
Sugar and tooth decay, 162
Egalitarianism, 50
Ehrenreich, Barbara, 274
Ehrlich, Paul and Anne, 52, 78, 216
Enclosure of commons, 30, 31
Enslavement. *See* Domestication
Epidemic disease, 227
Erosion. *See* Topsoil
Exotic
Exotic animals, 104
Exotic pathogens, 148, 179, 207, 212, 213
Exotic plants, 104, 212
Extinct animals, Britain, 26

F

Fagan, Brian, 176
Fairies, 45, 306
Fairyland, 46, 246, 258, 260, 266, 283, 306
Famine, 80, 104, 175
Fennians, ancient Finns, 43
Fertilizer
Bone meal, 133
Chemical, 108

Compost, 9, 139
Depleted soil. *See* Topsoil
Early soil treatments, 127
Fallowing, 109, 129
Fish in corn mounds, 184
Guano, 127, 133
Humanure, 8, 109, 127, 140, 141, 177
Manure, 104, 109, 140, 174
Mineral treatments, 128
Nitrogen, 128
Nutrients needed by plants, 126
Peak fertilizer, 142
Phosphorus (phosphate), 133
Potassium (potash), 134
River silt, 166, 170
Serious drawbacks, 137, 139
Toxic sludge, 137
Fish mining. *See* Mining
Flannery, Tim, 50
Fleming, Alexander, 231
Flour & health problems, 159
Forests, 92–98
Fortified villages, 42, 186
Freuchen, Peter, 250
Fungicide, 108, 151

G

Geese, 202
Genetically modified organisms (GMO). *See* Transgenic
Geronimo, 258
Ghost Dance, 257, 261
Giants, in old Europe, 45
Gilgamesh of Uruk, 94
Goebbels, Joseph, 254, 267
Good old days, 98
Grain
    Ecosystem destruction, 194
    Grain-fed meat, 195
    Stored grain is wealth & power, 84
    Versatile food, 76
    Wild grass seeds, 76, 206
Great Healing, 264, 265, 303
Great Leap Forward, 51–56
Green Corn Ceremony, 65
Green manure, 129
Green Revolution, 103, 149, 153, 154, 162, 180
Greenland collapse, 253
Greer, John Michael, 258
Greger, Michael, 227
Grimm, Jacob, 46

Grow or die, 82
Guinea pigs, 203
Gunther, John, 161, 197, 224, 266
Gwen of Cwmbelan, 296, 298

H

Haber-Bosch process, 130
Haiti, damaged ecosystem, 119
Harlan, Jack, 77
Healing
    Breaking historic habits, 60, 115, 263
    Do the best for the land, 298
    Eucatastrophe, 303
    Inner healing, 245–80
    Marriage to land, 281–84
    Restoring Fairyland, 46, 308
Heart disease, 158, 161
Heinberg, Richard, 106
Herbicide, 108, 151
Hesiod (Works and Days), 86
Hierarchy, need for, 83, 84, 186
Hillel, Daniel, 172
Hitler, Adolph, 239, 254, 255, 256
Hoarding not instinctual, 65
Hominids, origins of, 47, 66
Horse
    Changed Native American life, 74
    Domestication of, 73
    Domestication of horses, 76
    Manure for fertilizer, 140
    Need for pasture land, 103, 108
    Odin's horse Sleipnir, 74
    Used in farming, 35, 75, 173
    Used in logging, 75
    Used in warfare, 74, 91, 93
Humans
    Domestication of humans, 75
    Fatally flawed?, 57, 58
    Purpose of life, 101

I

Indigenous soul, 246, 274
Individualism, 48, 216, 240, 265
Industrial Revolution, 27, 37, 43, 96, 242
Infanticide, 215, 241, 242
Influenza, 33, 179, 180, 213, 228, 234
Inner knowing, 273
Insecticide, 151
Insects as food, 201
Instinctual mind, 250

Interspecies communication, 274
Iron Age, 26, 86, 90
Iron tools and weapons, 91
Irrigation
    Annual flooding, 166
    Dams, 121, 167
    Drip irrigation, 121, 122
    Flooded paddies, 177
    Future challenges, 124
    Groundwater mining, 177, 195, 199, 210
    Last 200 years, 103
    Limitations & problems, 120
    Rain-fed agriculture, 119
    Salinization of soil, 120, 122, 123, 124, 126, 163, 167, 170
    Water mining, 109, 120, 123, 124, 125, 170, 195, 199, 210
    Water wars, 126
    Waterlogging of soil, 120, 122, 123, 167
Ishmael by Daniel Quinn, 14
Isle of Rum, 31, 176
Isolation of modern life, 265

J

Jackson, Wes, 116, 174, 196, 204, 205, 217
Jefferson, Thomas, 86
Jenkins, Joseph, 141
Jewel and Hugo, 12
Jung, Carl, 247
Junge, Traudl, 256
Junk food, 157

K

Kalapuya tribe, 41, 305
Keith, Lierre, 195
Keweenaw ranch, 1–14, 305
King, F. H., 140
Krech, Shepard, 47, 187
Kung bushmen, 11, 56

L

Lakota tribe, 93
Lame Deer, 67, 70, 257
Lancashire riots, 28
Last great thrash, 111
Leonard, Annie, 267
Levy, Stuart B., 231
Liedloff, Jean, 249

Life on Earth begins, 45
Lime (calcium carbonate), 135
Livingston, John A., 58, 59, 217
Loess soil, 145, 168
Loneliness, 265
Long eyes, 41, 305
Louv, Richard, 280
Luddite Rebellion, 28

M

Maasai tribe, 71, 176, 197, 198
Machine-powered society, 263
Macy, Joanna, 279
Maize. *See* Corn
Malaria, 178, 179, 213, 230
Malthus, T. R., 234
Mann, Charles, 172, 185, 210
Manning, Richard, 118, 130, 153, 195, 199, 217
Margolin, Malcolm, 206
Marsh, George P., 178
Martin, Paul, 56
Marx, Karl, 236
Mayan civilization, 90, 181, 246
Measles, 42, 213, 230
Meat
    Agroforest-fed, 211
    CAFO, animals confined, 73, 196, 199
    Grain-fed meat, 195
    Grass-fed meat, 194
    Greenhouse gas emissions, 197
    Health effects, 197
    Herding not foolproof, 198
    Overgrazing, 72
    Wild meat is ideal, 199
Medieval Warm Period, 36
Megafauna extinction, 47, 56
Mennonites, soil mining, 173
Metal making, 46, 89, 90, 91
Michigan Tech (MTU), 14, 251
Micmac tribe of Maine, 271
Mining
    Copper mining, 21
    Fish mining, 21, 188, 239
    Forest mining, 21
    Fur mining (beaver), 21
    Iron mining, 21
    Mining ecosystems, 41, 225
    Resources are infinite, 22
    Water mining. *See* Irrigation
Misanthropy, 57
Mississippian culture, 182, 185

Mohawk tribe, 111
Molyneaux, Paul, 189
Montgomery, David, 217
Montgomery, Sy, 224
More is better, 81
Mowat, Farley, 274
Munch, Peter Andreas, 261
Muscle-powered society, 108, 175, 263

N

Naming trees of Shawnee, 282
Nature deficit disorder, 280
Navajo herding conflicts, 71
Necessities, reevaluating, 115, 266
New Guinea, 151, 171–73
   Population explosion, 173
Nile River, 122, 166
Nomadic foragers
   Benefitted from weapons, 55
   Damage to ecosystem, 52
   Lifestyle and beliefs, 48, 273
   Population management, 240
   Relatively sustainable, 47
   Safety cushion of food, 79
   Size of clans, 83
   Survive serious drought, 79
   Zero need for modern tech, 60
Norem, Julie K., 276
Norwegian ancestors, 38
Nuclear energy, 107

O

Oakland County, Michigan, 24, 34
Odin, Norse god, 74
Ohlone tribe, 41, 206, 219
Olive growing, 210
O'Mad, Henry, 112
One-Child Policy of China, 240
Ootek the wolf man, 274
Organic farming harms, 78, 166, 168, 171, 173, 175, 177, 181
Original mind. *See* Unconscious mind
Overgrazing, 72, 73
Overpopulation, 83, 100, 234, 237, 262

P

Paleopathology, 187
Peak everything

Era of cheap energy past, 107, 128, 131, 165, 179, 180, 259
Peak coal, 106
Peak fertilizer, 131, 134, 142
Peak food, 108, 115, 162
Peak natural gas, 106
Peak of eco-destruction?, 107
Peak oil, 105
Peak population, 237, 262
Perennial grains, 204
Perennial plants, 117
Pesticide, 108, 180
Photovoltaic power (PV), 7
Pigeons, 203
Pine root (turpentine fuel), 259
Plato, 95
Pleistocene overkill hypothesis, 56
Plow
   Benefits of tilling, 143
   Damages soil, 74, 116, 118, 120, 144, 145, 174
   Detested by fairies, 46
   Moldboard, 35, 91
Pollan, Michael, 157, 175
Ponting, Clive, 100, 101, 137, 228
Populate or perish, 82
Population explosion
   Corn farming, 186
   Europe, 36, 41, 42
   Green Revolution, 153
   Last 200 years, 104
   Nitrogen fertilizer, 129
Population management, 215, 240
Population reduction, 238, 262
Positive thinking, 274
Possessions, 63–66, 70, 84
Postel, Sandra, 120
Pot, Pol, 239
Potato
   Introduced in Norway, 39
   Potato blight, 80, 119, 149, 152, 213
   Potatoes & fungicides, 151
Potlatch ceremonies, 64
Poultry, non-industrial, 202
Power looms, 27
Prechtel, Martín, 90, 246
Predators, man-eating, 55, 224
Price, Weston, 157, 158, 197
Pritchard, Evan T., 272
Progress, myth of, 61, 236, 243, 255
Prosperity bubble, last 200 years, 100, 102, 105
Protect the Earth staff, 14

Protein foods, 199
Puritans vs. Indians, 75, 87
Pygmy, 48, 56, 241, 251

Q

Quincy Mine, Keweenaw, 292
Quinn, Daniel, 14, 298

R

Ragnarök, 260
Reader, John, 79
Reason, limitations of, 5, 18, 248
Red Cloud, 93
Reisner, Marc, 145
Repeated mistakes of civilization, 259, 263
Resistant weeds, pests, pathogens, 151–53, 231
Rice
    Diseases of farming, 142
    Diseases of rice farmers, 178
    Farms promote new flu strains, 179
    Few varieties planted, 150
    Flooded paddy farming, 177
    Methane from paddies, 178
    Pesticides poison farmers, 180
    Rice fungus, 80
Rift Valley fever, 122
Rinderpest, 80, 198, 213
Rixson, Denis, 176
Road and raid, same root word, 93
Roberts, Paul, 175, 179, 196
Rockefeller, Abby, 135
Rodenticide, 151
Roller mills, flour & sugar, 159

S

Salatin, Joel, 199
Salinization. *See* Irrigation
Salmon cultures, 26, 34, 38, 64, 136, 206
Sedentary living, 64, 77
Sewage treatment history, 135
Shepard, Paul, 67
Shoshone tribe, 65, 74
Simula, Vern, 14, 251
Sitting Bull, 258
Slash & burn. *See* Agriculture
Smallpox, 39, 42, 74, 213, 226, 230, 234
Smith, J. R., 211, 217
Sng'oi tribe, 273

Snorri Sturluson, 262
Soil erosion. *See* Topsoil
Speer, Albert, 255
Spellberg, Brad, 234
Spiritual economy of Mayans, 90
Stalin, Joseph, 104, 239
Steam engine, 27
Sugar & health problems, 159, 160, 162
Sustainable
    Sustainable agriculture?, 165, 204, 207, 213
    Sustainable logging?, 97
    Sustainable mining?, 90
    Wild cultures worked, 47, 78, 206, 218

T

Tacitus (Germania), 43, 94
Takelma tribe, 41
Tasaday tribe, 88
Tasaday, Philippine tribe, 89
Tauripan tribe, 250
Technology Fairy, 112, 237, 255, 259
Tent caterpillar defoliation, 208
Textile mills, 27
Three sisters farming, 181
Tikopia, 213
Tooth decay, 157, 161
Topsoil
    Community of micro-life, 115
    Creation of, 116
    Destruction via agriculture, 116
    Dust storms, 145
    Northern Europe, low erosion, 174
    Nutrient depletion, 80, 108, 127, 144, 172, 183, 185
    Soil mining, 78, 97, 117, 168, 169, 173, 185, 195, 211, 239
    US topsoil loss, 165, 211
    World topsoil loss, 116
Toxic sludge, 137
Trail of Tears, Cherokee, 88
Transgenic crop plants, 155
Trudell, John, 279
Tuberculosis, 158, 161, 187, 230
Tudge, Colin, 217
Turkeys, 202
Turnbull, Colin, 48, 250
Turpentine as biomass fuel, 259
Typhoid, 136, 142
Typhus, 36, 42, 230

U

Unconscious mind, 247
Unpeople, 31

V

Vanir (Norse gods), 260
Vegetarian impacts, 194
Ventura, Michael, 278
Venus figurines, 51
Vikings, 38, 40, 91, 94
Virus
    Animal diseases, 179, 234
    Aquaculture diseases, 189
    Cocoa trees, 207

W

Ward, Albert, 271
Waterwheels, 27
Wavoka the prophet, 257
We are all related, 45, 246
Weatherford, Jack, 251
Weatherwax, Paul, 185
West Bloomfield Township, 34
Wheat stem rust (UG99), 149
White Pine Mine, 14

Wild and free, the concept, 75, 272, 307
Willamette River, 305
Windmills, 27
Wisent, European buffalo, 26, 34, 73
Wolf communication, 274
Wolff, Robert, 273
Women
    Breast milk polluted, 112
    Family planning, 241
    Well treated in wild tribes, 49
Wood as a resource, 95, 259
Wood gasification for fuel, 259
Worldview, 195, 245–80
Wounded Knee, 258
Wright, Ronald, 217

Y

Yellow River, 122, 124, 140, 168
Yequana tribe, 250
Yoopers of the U.P., 21
Youngquist, Walter, 106, 120
Yurok tribe, 41

Z

Zedong, Mao, 51, 104
Zipsers, 36

ACKNOWLEDGEMENTS

I wish to extend thanks to those who have provided assistance to this project — Kate Alvord, Bairbre Flood, Slug Woman, Sandra Harting, Lisa Keller, Henry O'Mad, Kathy Russell, Vern Simula, Mary Waldner, Terry Shistar, Tom Warren, Walter Youngquist, and Wiktor Zelazny. Thanks so much!

This book is the result of a long pilgrimage in search of wisdom, and I spent many years exploring the minds of thousands of amazing wordsmiths, dreamers, healers, and visionaries. Since the purpose of this book is not scholarly, I chose to kill fewer trees by not providing hundreds and hundreds of footnotes. But I fondly tip my hat to the many people who have shared their finest ideas with the world. Thank you so much!

Acknowledgements

Made in the USA
Lexington, KY
30 September 2018